Historical Dictionary of Data Processing

Historical Dictionary of Data Processing

ORGANIZATIONS

James W. Cortada

Greenwood Press
NEW YORK
WESTPORT, CONNECTICUT
LONDON

Library of Congress Cataloging-in-Publication Data

Cortada, James W.
Historical dictionary of data processing—organizations.

Includes index.
1. Computer industry—History. 2. Computer service industry—History. 3. Electronic data processing—Societies, etc.—History. 4. Electronic data processing—Research—Societies, etc.—History.
I. Title.
HD9696.C62C47 1987 338.7′61004′09 86-19394
ISBN 0-313-23303-9 (lib. bdg. : alk. paper)

Library of Congress Catalog Card Number: 86-19394
ISBN: 0-313-23303-9

First published in 1987

Greenwood Press, Inc.
88 Post Road West, Westport, Connecticut 06881

Printed in the United States of America

The paper used in this book complies with the Permanent Paper Standard issued by the National Information Standards Organization (Z39.48-1984).

10 9 8 7 6 5 4 3 2 1

Contents

Preface

This historical dictionary is one of three companion volumes being published by Greenwood Press to provide basic yet important research tools for persons who have a general interest in the history of data processing. The three dictionaries will cover biographies (*Historical Dictionary of Data Processing: Biographies*); a range of scientific achievements and events encompassing computers, software, programming languages, and various devices and gadgets, old and new (*Historical Dictionary of Data Processing: Technology*); and organizations, such as companies, societies, and laboratories, that have been important components of the data processing industry (this volume, *Historical Dictionary of Data Processing: Organizations*). The dictionaries, taken together, will provide some 400 entries on all aspects of data processing, ranging from the beginnings to recent years. The attempt is to provide basic factual material and, where appropriate, historical interpretation to indicate the significance of specific facts. The entries conclude with bibliographic references to lead the reader to additional material.

These research guides can be used as if they were one. When an entry mentions a topic that has its own entry, a cross-reference is indicated by one of three symbols. The * indicates that the subject entry is in the dictionary devoted to technology; the ** points to the biographical dictionary; and † indicates the topic is in this volume.

Although indexes in each volume provide access to material within entries, the reader may feel that additional topics could have been included. Yet space limitations, economics, and the availability of reliable data made it necessary to limit the entries to those included. As the field of data processing history matures, an even more ambitious publication can be considered.

The history of data processing has come into its own since the mid–1970s and is now a respectable subject for scholarly examination. Thus, the introduction of these volumes is timely. Articles and books on the subject are appearing weekly. Like the industry itself, publishing on this theme has enjoyed explosive growth. Ten years ago, historical literature on data processing was obscure and meager, usually including reminiscences by scientists about computers built in the 1930s and 1940s. But that has changed. While working on my earlier bibliographic guide, I accumulated over 1,000 titles with little difficulty.[1] That bibliography barely tapped what was now a new field, a growth industry for publications. In some universities, students can take courses on the history of computing, and children (both young and adult) can visit computer exhibits in various American cities. For the historian, there is even a research organization dedicated to the history of computing—the Charles Babbage Institute (CBI).†

Of course, the industry has become extremely large and important. According to U.S. government and industry observers, the industry in the United States represented 3.4 percent of the gross national product in 1984—about the same as the oil industry and the automobile business. The latter two industries were expected to increase by 4 to 9 percent per year for the next decade, and data processing was expected to achieve some 6 percent of GNP by 1994.[2] In simple dollars, data processing was a trillion dollar industry worldwide in 1986. Since the industry was born and is headquartered in the United States, its influence on the economy has been greatest here. The data processing portion of the U.S. economy grew by over 10 percent compounded each year during the 1970s, and in the 1980s, it enjoyed several years when growth approached 15 percent. Put another way, in 1960 less than 10 percent of all workers used computers to perform their duties. By the mid–1980s well over 50 percent used them. U.S. Department of Labor studies suggest that dependence on computerized technologies will climb to over 70 percent by the end of the decade. One International Business Machines Corporation (IBM)† executive predicted, in 1985, that by 1994 one out of every two workers will have a terminal. In 1984 five out of a hundred had one.[3]

Historically, forecasts about the growth of data processing have been too conservative. In the late 1940s, for example, many vendors thought worldwide demand for computers might be twenty to forty machines. When IBM announced the S/360* in 1964, the company expected to sell several thousand. By 1968, 15 percent of all computers worldwide were S/360s, and the company had doubled its revenues in less than five years. If data processing permeates society with the same intensity that electricity, radio and television, and the ubiquitous automobile have, one might reasonably conclude that forecasts for growth over the next two decades are shy. The second half of the twentieth century is not the age of atomic energy, as expected after World War II, but the age of information, or a postindustrial society. Indeed, according to James Martin, a respected industry leader, a typical company in the industrialized world in the early 1980s was spending nearly 30 percent of its budget on handling information—telephone calls, letter writing, talking, publications, and data processing. In other words, over 3 percent of an organization's budget was being spent on data processing (machine readable data). Information had become as important to a company's health and raison d'être as inventory, people, or cash.[4]

Interest in the history of data processing is thus timely. With that interest comes the need to address certain kinds of questions of concern to historians—issues which involve identifying the origins of modern data processing, the role played by specific people, the effects of technologies, the role of information systems and their acceptance and use, and, finally, the impact of organizations in developing, marketing, installing, and using computers. Historians have begun the difficult process of identifying the fundamental influence that data processing has had on society.[5] Some scholars are taking the initial steps by chronicling the major events in the industry's first forty to fifty years. Considerable spade work remains to be done on the origins of modern data processing and computers. That spade work inevitably must lead researchers through a thicket of tabulating equipment back to the introduction of printed records, engines, and other mechanical aids for handling information.[6]

For those first, halting steps in the formal history of data processing, the necessary research tools are almost totally lacking. The existence of the journal *Annals of the History of Computing*, first published in 1979, was a major thrust in the right direction; so too was the founding of the CBI in 1977. Increased attention came with the establishment of professorships for the history of science and technology in general, and, in some cases, of data processing. Many of the computer "pioneers" of the 1940s and 1950s actively sought to preserve information about the early days as well.[7] Finally, the willingness of journals and book publishers to carry material on the subject has also spurred activity in this field.

This particular volume represents part of this new historical movement. Although the data processing industry is highly visible, its components and mode of operation are little understood. This historical dictionary attempts to address related issues. It includes entries on companies that were important historically, on societies and associations of data processing professionals and scientists, and on laboratories and various government agencies. An introduction on the data processing industry provides an overall view of the topic. The entries suggest important issues as reflected in current historical literature and point to available bibliography (of which there is very little concerning institutions and companies). This dictionary is not intended to be a definitive history of data processing or the ultimate guide for the specialist. For the benefit of the general user and students, major subjects are reviewed with the assumption that the reader has little knowledge of data processing technology. For the specialist and scholar, the entries provide a starting point and a ready reference as well as details on technological and economic issues. In short, an attempt has been made to provide a balance between the needs of the general reader and the specialist. Each entry describes the significance of the topic, along with the fundamental facts as currently available and understood.

The criteria for selecting organizations to include in this volume were very specific. First, my intent was to include organizations overwhelmingly American in origin and significance, because the bulk of the data processing industry has existed in the U.S. Hence the reader will see very few organizations from Great Britain or Japan as examples; they have been included when their historic influence within the U.S. industry was viewed as obvious. Second, institutions which have been recognized by historians as having already made significant contributions to the data processing industry—IBM and Apple Computers, for example—were included. Third, organizations which no longer exist but have become parts of others have been added. These include Remington Rand,† which became part of Sperry,† both of which produced and marketed the UNIVAC* computers of the 1950s. In total, 81 organizations are reviewed in this volume. In my opinion, these constitute the most important in the history of the American data processing industry.

The first edition represents an initial attempt to gather historical data in an encyclopedic fashion. I welcome suggestions for improving this or any of the companion volumes. Yet I do not want to shift blame for views and errors found in this volume on others. I am solely responsible for the views and contents of this book, not my employer, friends, or others who patiently tried to make this project more than I dreamed it could be in the beginning.

NOTES

1. James W. Cortada, *An Annotated Bibliography on the History of Data Processing* (Westport, Conn.: Greenwood Press, 1983).

2. Peter L. Schavoir, "The Industry: Future Without Limits," *Digest*, vol. 4, no. 1 (January/February 1985): 9–11.

3. Ibid.

4. James W. Cortada, *Strategic Data Processing: Considerations for Management* (Englewood Cliffs, N.J.: Prentice-Hall, 1984): 4–26; James Martin, *Application Development Without Programmers* (Englewood Cliffs, N.J.: Prentice-Hall, 1982): 1–13.

5. Good examples of this concern include Hiroshi Inose and John R. Pierce, *Information Technology and Civilization* (New York: W. H. Freeman and Co., 1984); Stanley Rothman and Charles Mosmann, *Computers and Society* (Chicago: Science Research Associates, 1976).

6. At the dawn of modern data processing, these issues were already obvious. See Norbert Wiener, *Cybernetics: Or Control and Communication in the Animal and the Machine* (Cambridge, Mass.: MIT Press, 1948, 1961) and a series of lectures by the father of the modern computer, John von Neumann, *The Computer and the Brain* (New Haven, Conn.: Yale University Press, 1958). Two recent studies are by Walter M. Mathews, ed., *Monster or Messiah? The Computer's Impact on Society* (Jackson: University of Mississippi Press, 1980); Sherry Turkle, *The Second Self: Computers and the Human Spirit* (New York: Simon & Schuster, 1984).

7. Two representative collections of such memoirs are N. Metropolis et al., eds., *A History of Computing in the Twentieth Century* (New York: Academic Press, 1980); Richard L. Wexelblat, ed., *History of Programming Languages* (New York: Academic Press, 1981).

Acknowledgments

This project would not have been possible without the encouragement, suggestions, patience, and knowledge of many people. Librarians at Vassar College, the U.S. Library of Congress, the Smithsonian Institution, University of Virginia, Vanderbilt University, and IBM's library in Poughkeepsie, New York, all made materials available to me. To each of them I owe an enormous debt of gratitude.

So many individuals helped that it would be difficult to mention them all. However, a few deserve special recognition. Hank Tropp suggested many entries, gave me leads to information, and served as all-round encourager throughout the project. Nancy Stern edited my original list of proposed entries and probably wondered if they would ever be written. But she was supportive as I launched into the three books. Paul Ceruzzi was also supportive. Just as exciting was the pleasure and honor of talking to living "pioneers," such as the wonderful Mina Rees, as I chased down information. Jean E. Sammet set a high standard of scholarship along with Emerson Pugh. The editor of the *Annals*, Bernard A. Galler, cheered me on in the beginning.

At Greenwood Press, home for nearly a half dozen of my earlier projects, friends and professionals were patient with me as I took too long to write these volumes and then agreed to publish three rather than one book. Cynthia Harris, acquisition editor at Greenwood, was both an outstanding manager and an intellectual asset. Betty Pessagno copyedited the manuscript with great care. The Greenwood staff, always efficient and effective, moved this book through to publication. I would like to thank those publishers who allowed me to use material quoted or otherwise used in this dictionary.

A special thanks goes to my family for living around my notes, appreciating their importance to me without necessarily understanding them. That was particularly true of my wife Dora who managed to run a house while I wrote the book. Everyone who helped showed the kind of faith and enthusiasm that have been characteristic of the data processing industry since its inception.

Historical Dictionary of Data Processing

An Introduction to the History of the Data Processing Industry

The data processing industry did not come into existence until the late 1940s, when digital electronic computers* were first constructed. The use of equipment to process data had existed much longer, however. The issue of when the sellers and users of such devices finally constituted an industry is difficult to pinpoint. A collection of similar economic activities can be described as an industry when a series of services or products lend themselves to a collective identity. An industry is also characterized by a set of dependencies among its members. Thus, for example, a dependency might include makers of components that go into a computer, then the actual manufacturer, and finally the data center that uses it. The act of identification is initially the result of the recognition that certain efforts represent a common set of actions. In the next phase economists or historians name a cluster of economic entities as an industry.

Looking at data processing this way, the era of modern data processing (that is, based on the use of digital computers) began during the 1940s, although historical antecedents existed. The predata processing era can be defined as any time prior to the implementation of modern punched card technologies in the 1880s. The use of these cards marked the transition between an era when little technology was employed to manipulate data to one that relied increasingly on the use of electronic components. The great era of punched cards lasted from the 1880s through the 1940s, with almost no competition from other input/output gear, and during the 1950s punched cards rivaled such electronic technologies before declining to minor status in the 1960s.

The lessons learned before the dawn of modern data processing influenced the early development of the new industry. For example, the gathering and manipulation of data had been ongoing for centuries. Scribes and artists left humankind's written records on the walls of caves. At the dawn of recorded history (circa 10,000 B.C.) languages together with alphabets came into being, and writing became a more formal discipline. These developments displaced human memory as the primary repository of information, allowing vast quantities of data to be gathered and used. The ancient Egyptians improved writing instruments, with their use of papyrus signaling a new level of technological innovation. China's introduction of paper allowed new levels of price/performance in the handling of data, and the development of the movable printing press during the Renaissance rapidly and radically changed information management. In net terms, the amount of information that could be recorded and used exploded exponentially within two generations. The number of people who could now access data also experienced similar expansion. Books fundamentally altered the very nature of thinking and attitudes.

These changes directly paralleled the history of data processing in the twentieth century. For example, like books computers allowed greater amounts of information to be gathered, stored, manipulated, and ultimately used. The role of pencils, paper, and books was very instructive to historians who were studying data processing use patterns and the sociological and philosophical implications of such technologies. The most important characteristic of the book relevant to that of data processing, for example, was the increase in the amount of information that became available. Books were easier and quicker to produce and less expensive than hand-copied manuscripts. Information became a symbol and source of power in society, unleashing economic and political forces that helped speed up the rate of change in the Western world. Information (some historians use the term *knowledge*) molded a society and economy that became increasingly dependent on information. As a byproduct of the Industrial Revolution, increased pressure grew for better management of more information. The correlation between major changes in society and the growth in information was direct. Historians should give more attention to this correlation, particularly in relation to events in the past five centuries.

In the 1800s the most important impetus for better data management came from the need to manipulate increasing quantities of numbers. By the early 1800s, for example, navigational tables were being corrected and enhanced, encouraging such individuals as Charles Babbage** to develop machinery to do the work. During the mid–1800s insurance companies, railroads, and governments began the manual collection of large amounts of data at enormous costs. The need to speed up this process, improve accuracy, and reduce costs became the bedrock of technological innovations in the late 1800s. By the 1880s technology had entered the office. Pencils and pens made their appearance, along with adding machines and cash registers. The telegraph, which had existed since before the American Civil War, was used to transmit information in ever increasing amounts, and, of course, the telephone made its initial appearance in the 1870s. New printing technologies were developed throughout the century, causing publishing costs to drop and literacy rates to rise. By the end of the century, the typewriter had become a common instrument. The technology rate was higher in American than European organizations. For example, in 1919 at the end of World War I only Washington's diplomatic delegation used typewriters.

By 1880 the belief was widespread that mechanical devices could handle many traditional office tasks. By 1890 that attitude was also dominant in factories. Thus, the time was ripe for changes in the application of technology to the manipulation of data. At this point card punch equipment was introduced. Herman Hollerith's** collection of tabulating equipment created a new industry by the early twentieth century. This new industry was variously called office products, office automation, office equipment, tabulating industry, and the card punch business, but by the end of World War I an identifiable sector of the economy used punched cards and related equipment to manipulate information. Increased reliance on such technology during the 1920s and 1930s motivated many to graduate to more powerful technologies in the 1940s.

The growing demand for better ways to manipulate data on cards coincided with the development of what would become computers in the late 1930s and early 1940s. After World War II, one would think that the technology developed for war-related projects would next be applied to the civilian manipulation of information. The facts proved otherwise. In the mid–1940s there was concern about whether the market was large enough to justify the investments necessary to develop computers. Questions concerning demand for such technology were raised at International Business Machines Corporation (IBM)† and National Cash Register Company (NCR)†, as well as within well-established electronics firms. Only those who had worked on government projects during World War II and were close to the early devices showed faith in computers. This group took the first steps to build commercial machines and to establish data processing companies. By 1950 the more established firms concluded that data processing equipment could profitably displace card-tabulating and sorting products. The use of data processing increased dramatically during the 1960s as a result of sharply declining costs for computing, better technology, and increased ease of use. By the 1970s computer technology was an important influence on the industrialized world's economy and, by the 1980s, on the culture of some nations, especially the United States.

The following discussion centers on the United States because most developments in the data processing industry have taken place there.

Modern data processing is rooted in card punch technology as a method for recording and manipulating data. Although its origin is traditionally ascribed to the work of Herman Hollerith in the late 1800s, in fact, various Americans and Europeans had been developing cards with punched holes and equipment to trap or manipulate information since the early 1700s. It was not until the late 1870s, however, that the use of cards was sufficiently widespread to help launch a new set of economic opportunities, let alone an industry. Hollerith came to Washington, D.C., in 1879 to begin work at the U.S. Bureau of the Census to help with the following year's count of the nation's population. The census was taken primarily without the help of data-handling equipment. The information collected that year was barely analyzed and final reports completed before it was time to take another census in 1890. It was becoming obvious that with the United States' phenomenal growth in population more advanced census methods would be required.

In 1884 Hollerith obtained patents for equipment to punch and read holes in cards, each hole representing one piece of information. The punched cards—which would remain in use for over 100 years—provided a portable, yet standard, medium for interchanging and recording data. Thus, information could be processed and shuffled into various categories, enabling a wide variety of tabulating applications to emerge over the next half-century. Hollerith also used electrical power for what otherwise were Jacquard**-like devices of the past to process these cards.

During the 1880s Hollerith convinced various organizations to use his equipment in gathering statistics. The city of Baltimore, for example, used his devices in 1886 to collect data on health conditions, making this government perhaps the first user of punched card equipment in data processing. Hollerith's major breakthrough came when the U.S. Bureau of the Census adopted his equipment for the Census of 1890. It was to be a complicated census, for the agency now planned to obtain information on 235 topics per person versus 215 in 1880 and only 5 in 1870. Furthermore, the population had grown, requiring both greater and quicker number counts than in 1880. His equipment proved a complete success. On December 12, 1890, the government announced that the U.S. population totaled 62,622,250, and it completed its analyses of various data elements within a couple of years. Hollerith's equipment saved the government over $5 million in expenses. During the 1890s governments in Russia, France, and several other countries adopted Hollerith's equipment to conduct their census studies of population and agriculture. Railroad companies and insurance firms were also persuaded to rent this gear and to buy cards from Hollerith's company. In short, by the early 1900s Hollerith and several other vendors (fewer than four) were operating tabulating companies in the United States and in Europe. And as would happen in the 1940s and 1950s various government agencies were sponsoring research and development of improved versions of such technology.

Hollerith's successes suggest how greatly the industry had grown. In 1896 he established the Tabulating Machine Company capitalized at $1 million. For the Census of 1900 he rented 311 tabulating machines, 20 automatic sorters, and 1,021 punches to the federal government. Revenues that year exceeded a half million dollars. In 1911 the company had assets of over $2.3 million. That year it was reorganized into C-T-R and sold. In 1914 the firm acquired a new general manager, Thomas J. Watson.** It was he who converted the little company into IBM.

In 1912 net profits for the firm reached $541,000, with equipment installed in both commercial and government accounts. During World War I, as in all twentieth-century wars, the need for data processing increased sharply and com-

Table 1
Revenues for Selected Business Machine Companies, 1928 (Millions of Dollars)

Company	Revenues	Earnings
Burroughs	32.1	8.3
International Business Machines	19.7	5.3
National Cash Register	49.0	7.8
Remington Rand	59.6	6.0
Underwood Elliott Fisher	19.0	4.9

SOURCE: *Moody's Industrial Manual*, 1930. The table is a modification of one from Robert Sobel, *IBM: Colossus in Transition* (New York: Times Books, 1981): 75. Copyright © 1981. Reprinted by permission.

panies that sold data-handling equipment did well. C-T-R had sales of $4.2 million in 1914 and, in 1917, $8.3 million, with earnings that year approaching $1.6 million. In 1918 the company was renting 1,400 tabulators and 1,000 sorters scattered across 650 accounts. The largest customers were still railroad and insurance companies and government agencies. It had no serious competitor, and thus the industry's size can be measured by studying C-T-R.

The 1920s held great promise of prosperity for firms in the office equipment market, as it was then known. That market defined itself as including typewriters, adding machines (such as NCR), and tabulating gear (such as IBM). Although the decade started with an economic recession that hurt C-T-R, the environment seemed favorable for the long term. The gross national product of the United States went from $69 billion in 1921 to $103 billion in 1929. Per capita income also rose, increasing from $641 in 1921 to $847 in 1929, fulfilling the promise of the little industry at the start of the decade. During the 1920s Americans confirmed the faith in new gadgets and machines, buying radios, refrigerators, and cars in record numbers. America became more urbanized and talked on more phones than ever before. The demand for typewriters, accounting devices, bookkeeping equipment, and tabulating gear rose steadily during the decade until the Great Depression.

IBM typifies the pattern of behavior within the office equipment industry during this era. In 1922 IBM's revenues totaled $10.7 million and hit a high of $20.3 million in 1931. During the 1930s revenues declined (as did those of most companies) to $31.7 million in 1937, yet closed out the decade with $46.3 million. Other major suppliers during the interwar period included Burroughs Corporation† and NCR. Remington Rand† remained a dominant presence in the office arena, offering a broad range of products, primarily adding machines and typewriters. Other major firms included the aging Powers Accounting Machine Corporation†—Hollerith's great rival in the early 1900s—and Underwood Elliott Fisher. As Table 1 suggests, these few companies dominated the market and

Table 2
Relationship of Card Sales to Total Sales, IBM, 1929–1934 (Millions of Dollars)

Year	Card Sales	Total Sales
1929	3.6	10.7
1930	3.9	9.9
1931	3.4	8.8
1932	2.8	7.1
1933	2.8	7.1
1934	3.4	8.7

SOURCE: *Moody's Industrial Manual*, 1932, 1936; and a table of which the above is a slight modification, Robert Sobel, *IBM: Colossus in Transition* (New York: Times Books, 1981): 84. Copyright © 1981. Reprinted by permission.

accounted for about $180 million in business by the end of 1928, the lion's share of the total.

The sale of cards also suggests the size and scope of the industry. IBM dominated in this area. By the mid–1930s the firm sold over 4 million cards per year. At a price of $1.05 per thousand, they generated an income of $4.2 million, which remained fairly steady throughout the decade. Yet in 1938 cards generated $5 million in revenue out of the company's total of $34.7 million. Throughout the 1930s the demand for machine-readable data increased, particularly within the New Deal government agencies. The largest user was the Social Security Administration. IBM's rental income from keypunches (used to convert data into holes in cards) grew from $743,000 in 1929 to $1.4 million in 1931. In 1933, the first year of the new Roosevelt administration, revenues climbed to $1.8 million. Thus, when these revenues along with the role of cards (see Table 2), are considered, the need for information grew despite the Depression. The economy was shifting rapidly from agriculture to manufacturing, from farm labor to office work—the natural market for information processing technology. IBM ended the decade as the largest supplier of office equipment. Yet the industry as a whole was still dominated by the same companies (Table 3).

World War II had a profound impact on the data processing industry. The war was the leading cause of the conversion from office tabulating equipment to data processing. All electronics firms, as well as office equipment manufacturers, devoted their attention to war-related projects. Some firms, such as IBM, made rifles and tabulating equipment for the military. Others helped develop sophisticated electronics to handle range-firing equipment and radar.

Most of this research, however, was done at American universities. One important activity involved the use of electronics at universities to build large calculating devices to prepare tables for ballistics, such as artillery range-firing

Table 3
Comparative Statistics for Selected Business Machine Companies in 1939 (Millions of Dollars)

Company	Revenues	Earnings
Burroughs	32.5	2.9
International Business Machines	39.5	9.1
National Cash Register	37.1	3.1
Remington Rand	43.4	1.6
Underwood Elliott Fisher	24.1	1.9

SOURCE: *Moody's Industrial Manual*, 1940; table from Robert Sobel, *IBM: Colossus in Transition* (New York: Times Books, 1981): 87. Copyright © 1981. Reprinted by permission.

charts, and to do calculations essential for the development of the atomic bomb. Major projects that had begun before the war (such as at MIT) or during the fighting for specific military purposes (as at the Moore School of Electrical Engineering† at the University of Pennsylvania) led to the construction of the first digital computers* by 1944 and 1945. All of these machines were unique. They proved so successful that government agencies ordered more of them following the war. Among these agencies were the military services and almost every major branch of government. The developments of World War II were not exclusively an American story. Others, particularly in Great Britain, built equipment designed to break enemy communication codes. As a consequence, Britain's knowledge of computational equipment was quite high at the end of the war. Yet the dominant field of activity in the area of computers remained overwhelmingly in the United States.

The industry grew on the basis of that expertise. On the one hand, American government agencies were willing to buy computers and to support research leading to their construction. On the other, individuals who had worked on such devices began to establish their own firms to satisfy that demand. Finally, the traditional electronics firms such as General Electric (GE),† and major office equipment vendors, such as IBM, had the potential to participate from the start in what would become the data processing industry.

Events moved very quickly in the late 1940s as government-sponsored research at companies and universities progressed, resulting in the development of new computers from a variety of sources by 1950. Machines were constructed at over twenty locations at U.S. universities and research institutes. In Great Britain nearly ten such projects were underway. Among the American projects were those at the Universities of California at Berkeley and Los Angeles, Harvard, University of Michigan, University of Pennsylvania (home of the ENIAC,* the first electronic digital computer), the Institute for Advanced Study, Los Alamos Scientific Laboratory, the National Bureau of Standards,† the U.S. Naval Research Laboratory, and the RAND Corporation. Thus, by the early 1950s considerable expertise existed on how to make workable stored-program electronic digital processors. Thus, those with technical knowledge and ability to obtain a contract from the U.S. government were potential participants in the nascent data processing industry.

Demand for these expensive pieces of equipment grew sharply during the early 1950s, far in excess of what had been experienced in the late 1940s. Yet in the beginning the primary customers were universities and government agencies. While a number of firms took advantage of the growing demand for computational devices, at the end of the 1950s the large office equipment manufacturing firm typically controlled the market. Also participating were some electronics companies. New firms, that appeared in the 1940s and early 1950s included the Engineering Research Associates (ERA)† and the Eckert-Mauchly Company (which later became Remington Rand,† the makers of the UNIVAC*). IBM moved into computers by the early 1950s along with Burroughs, NCR, and Remington Rand.

By the 1960s, IBM was struggling with the issue of whether or not to relinquish its hold on the tabulating market in favor of computers. It, together with others with a stake in an existing market, finally opted to lead customers from card punch gear to electronically based calculators in the late 1940s and early 1950s, to stored program devices (computers) that relied heavily on card input/output. In time, this equipment was replaced by new generations of computers that provided greater efficiency, more performance and capacity, and always at lower prices.

The general term used for data processing in the 1970s was the *computer industry*. In the mid–1960s the phrase *electronic data processing industry* had gained currency. The term *computer industry* more aptly described the new world of data processing during the 1950s. This industry was made up of organizations that either manufactured computers or provided computer-related services and supplies. Throughout the 1950s equipment was consciously developed which was directed at satisfying the need for more information in machine-readable form. Computers were at the center of the new industry.

It is not difficult to understand the economic and social bases for this phenomenon. The traditional shift in the workforce in America offers a quick insight. Between 1880 (when Hollerith was busy inventing tabulating equipment) and 1955, the percentage of the total U.S. workforce in agriculture decreased from approximately 45 percent to 37 percent. By 1980 it was below 25 percent. The industrial workforce grew in the late 1800s and then declined as a percentage of the total working population in the same period. Thus, that sector in 1880 accounted for about 27 to 28 percent of all workers but by 1955 had dropped below 15 percent and, by 1980, to less than half that share. The most important change apparently occurred in the service sector which owned just over 25 percent of the workforce from 1880 to World War II, over 35 percent in 1955, and nearly 50 percent in 1980.

Equally impressive were the trends in the information sector of the economy. In 1880 that group comprised less than 10 percent of the workforce, grew steadily in proportion over the next sixty years to nearly 25 percent in the late 1940s, and experienced a modest decline in the 1940s and early 1950s as the service sector grew sharply. Then it experienced a steady, even exponential, growth beginning in the second half of the 1950s. It passed the 30 percent mark by the early 1980s. Although many definitions were developed to define the information-based workforce, thereby resulting in different statistics, the trends were consistent. They all showed that the office-bound worker had become an increasingly prevalent force in the economy and had become more dependent on handling data and information. In other words, the percentage of their work requiring manipulation of data had risen.

The underlying reasons for these changes whereby the economy shifted from one that first grew food, then made goods, and finally manipulated information will be argued for decades to come. However, some incentives for change are readily discernible. For one thing, as a managerial class rose within the U.S. economy in the 1840s, becoming the "visible hand" so aptly described by Alfred D. Chandler, they needed information with which to manage companies. Recorded data proved more valuable than memory or experience. This was particularly the case after the amount of information needed to operate successfully an organization grew.

Second, the cost of gathering, manipulating, and using data through mechanical and electronic means kept declining over the preceding 100 years. Nowhere has this trend been more evident than with computers. For example, multiplying two numbers on a computer cost nearly $1.50 in the early 1950s but declined to a miniscule fraction of a penny by the late 1970s. Such decreases made the use of computers practical in business, government, universities, and many other applications. By the 1980s commentators on industrial society had concluded that data processing had taken on a life of its own that transcended simple economics, becoming a fundamental feature of developed societies.

The definition of the data processing industry remained relatively uncontroversial and constant from the 1950s until at least the early 1980s. Collectively and individually, participants in data processing defined industry participants as those who manufactured, sold, and serviced computers, peripheral equipment, software, and related services. By the early 1960s the growing community of users of computers no longer thought of themselves as simply scientists or engineers but as members of the data processing world. This phenomenon can be traced by studying the growth of various associations within the data processing community during that decade, such as the Data Processing Management Association (DPMA),† the Association for Computing Machinery (ACM),* and the American Federation of Information Processing Societies (AFIPS).*

Measuring the revenues of data processing suppliers suggests the extent to which the data processing industry had grown. As Table 4 illustrates, the industry was worth just less than $1 billion in the mid–1950s but climbed to over $28

Table 4
Data Processing Industry Shipments and Revenues by U.S. Firms, 1955–1979, Selected Years (Billions of Dollars)

Year	Domestic	Worldwide
1955	less than 1	less than 1
1960	0.5	1.0
1965	2.0	2.5
1970	7.0	10.0
1975	15.0	21.0
1979	28.0	37.0

Data represent estimations only, with a reliability variance of approximately 10 percent.

SOURCE: Montgomery Phister, Jr., "Computer Industry," in Anthony Ralston and Edwin D. Reilly, Jr., eds., *Encyclopedia of Computer Science and Engineering* (New York: Van Nostrand Reinhold, 1983): 335.

billion by the late 1970s. A similar trend was evident for the data processing industry worldwide. Thus, it stood at over $40 billion by the end of the 1970s. These statistics account for only the total shipment of products and services. Historically, these dollars represented between 20 and 40 percent of all expenditures for data processing. The other costs went toward salaries for users of data processing products or for the facilities in which they were housed. Furthermore, the above data reflect only the earnings of American companies. They do not take into account the activities of European firms which, although small, grew during the 1970s and early 1980s.

Over 50 percent of all revenues for the period (1952–1982) came from hardware. Percentages changed over time, however. Thus, for example, in 1955 over 80 percent of all shipments of products by U.S. firms (as measured in dollars) went toward hardware. That percentage dipped into the mid–70s percentile during the 1960s and into the 60s range during the 1970s. It leveled to about 50 percent by the early 1980s. During the same period, the percentage of contribution to the hardware numbers by computers themselves dropped even more sharply, especially after 1970 as expenditures for peripheral equipment rose along with the demand for more machine-readable data. Although these percentages shifted, the actual total amount of money spent on data processing increased. The use of more peripheral equipment per data system (each included at least one computer) became a boom subset of the data processing world because between 1968 and 1978, for example, those who sold such equipment grew on average by 25 percent annually.

Several other trends should be noted. First, the number of computer systems in use within the United States grew slowly during the ten years from 1955 to 1965 and then exponentially from 1965 to the present. Measuring the total number

Table 5
Value of Computer Systems in the United States, 1955–1979, Selected Years (Billions of Dollars)

Year	General Purpose	Minis	Small Business	Total
1955	0	0	0	0.5
1960	0	0	0	1.0
1965	5.0	0	0	5.0
1970	18.0	1.0	0	24.0
1975	30.0	5.0	0	35.0
1979	50.0	7.0	2.5	59.5

SOURCE: Estimated figures extracted from Figure 4, Montgomery Phister, Jr., "Computer Industry," in Anthony Ralston and Edwin D. Reilly, Jr., eds., *Encyclopedia of Computer Science and Engineering* (New York: Van Nostrand Reinhold, 1983): 337.

of computers in use by the thousands, there were fewer than 1,000 at the start of the 1950s, 100,000 in the early 1970s, nearly 300,000 by 1975, and beyond 450,000 at the end of the 1970s. Large mainframes drove the number of dollars expended up until the mid–1970s when their absolute number remained steady. What changed was the fact that general processors individually grew in size and value, accounting for the continued surge in the overall expenditure of dollars on computers since 1965. Measuring the total value of computer systems over time suggests the extent of the phenomenon (Table 5).

Hardware accounted for many dollars within the industry and obviously was the dominant topic of discussion concerning data processing. In the 1950s the single most expensive component of a system was the computer. The mainframe's percentage of dominance, measured in dollars, declined steadily after the 1960s, while other devices, disk drives, and terminals began consuming an ever larger portion of money spent on hardware. As Table 6 and Figures 1 and 2 show, the value of mainframes in thousands of dollars declined in the 1950s and 1960s and rose in absolute terms in the late 1970s. Terminals rose steadily along with the cost of peripherals during the entire period. As the price per byte of memory improved, their relative value in absolute dollars remained flat, with very moderate increases over time.

The extensive growth in expenditures and, consequently, values of peripherals over the past thirty years was dramatic. Online information storage sharply rose in value as the cost per byte declined. The amount of online storage nearly doubled each year from 1974 to the present. The total amount spent on tape

Table 6
Relative Price Performance for Selected IBM Computers, 1953–1979

IBM Machine	Relative Speed	Relative Price[a]	Relative Speed/Price
650 (1953)	1	1	1
360/30 (1964)	43	0.025	1,700
370/135 (1971)	214	0.011	19,000
4341 (1979)	1,143	0.001	1,143,000

a. The cost to execute one instruction per second.

Improvements in the performance/price ratio when speed is taken into account (expressed as relative speed divided by relative price) for some IBM machines.

SOURCE: R. Moreau, *The Computer Comes of Age: The People, the Hardware, and the Software* (Cambridge, Mass.: MIT Press, 1984): 191.

Figure 1
Relative Price Performance for Selected IBM Computers, 1931–1979

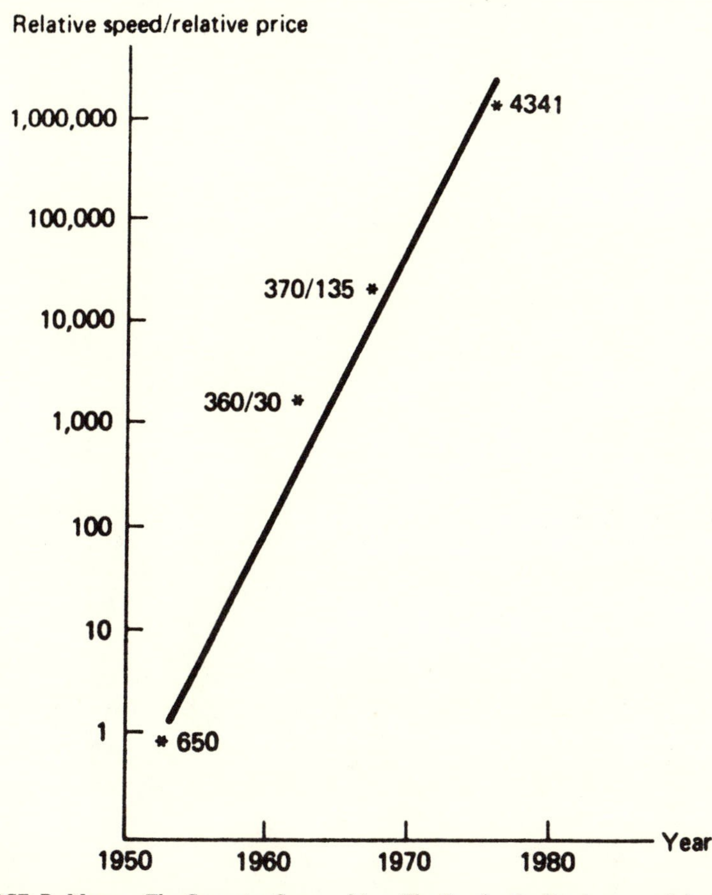

SOURCE: R. Moreau, *The Computer Comes of Age: The People, the Hardware, and the Software* (Cambridge, Mass.: MIT Press, 1984): 191.

drives declined as disks replaced them. Printers were very expensive when they first appeared, dropped in cost by nearly half during the 1960s and early 1970s, and then rose again as impact printers were slowly replaced with laser-based technologies that could handle hundreds of times more volume than earlier machines. As usage of terminals and magnetic storage media for input/output equipment rose, beginning in the early 1960s, the value of punched card equipment declined gently over the decades (see Figure 3). Throughout most of this period, data entry equipment never exceeded 5 percent of the overall value of computing gear, whereas computers always took more than half.

Figure 2
Components of an Average GP System

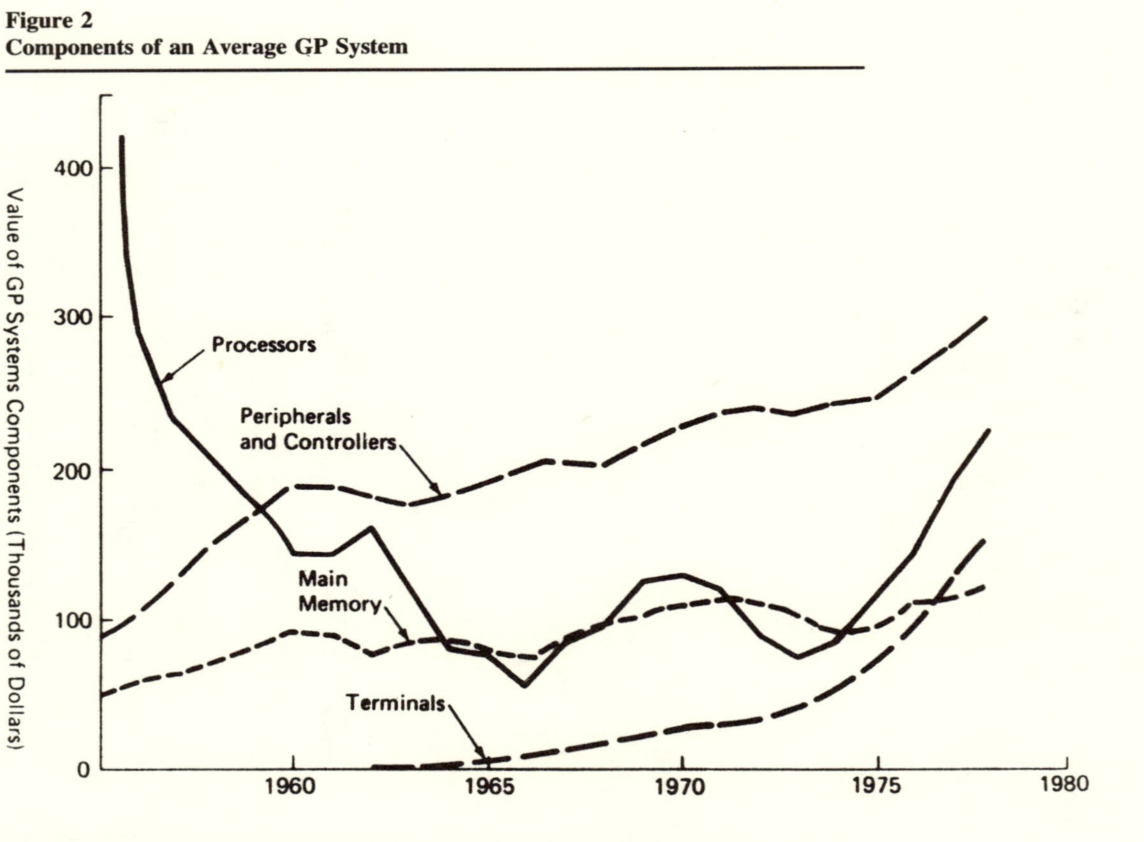

SOURCE: Montgomery Phister, Jr., "Computer Industry," in Anthony Ralston and Edwin D. Reilly, Jr., eds., *Encyclopedia of Computer Science and Engineering* (New York: Van Nostrand Reinhold, 1983): 338.

Figure 3
Total Value of Data Entry Equipment in Use in the United States and Its Percent Distribution

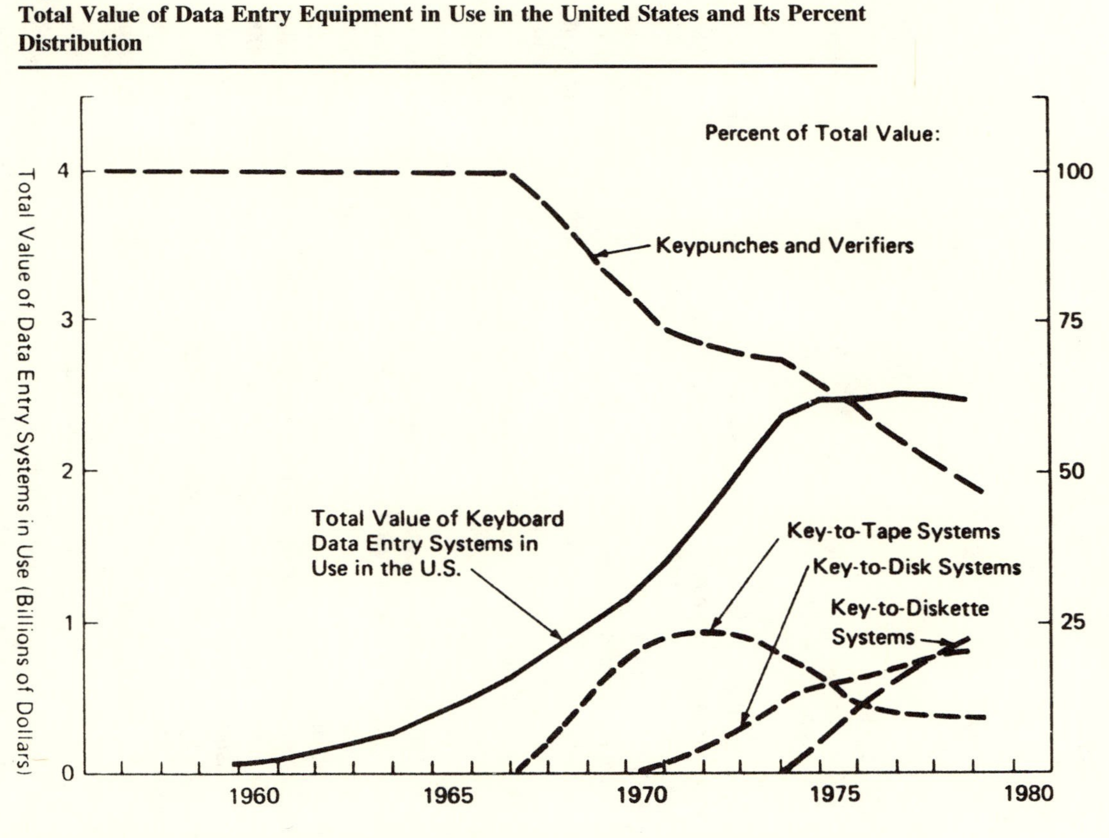

SOURCE: Montgomery Phister, Jr., "Computer Industry," in Anthony Ralston and Edwin D. Reilly, Jr., eds., *Encyclopedia of Computer Science and Engineering* (New York: Van Nostrand Reinhold, 1983): 339.

Table 7
Cost of Processing One Million Instructions, 1955–1981, Selected Years

Year	$ Cost/Million Instructions
1955	40.00
1961	2.00
1965	0.40
1971	0.11
1977	0.08
1979	0.04
1981	0.02

SOURCE: Data based on IBM computers. For details, see James W. Cortada, *Managing DP Hardware* (Englewood Cliffs, N.J.: Prentice-Hall, 1983): 14–21.

Although the history of digital computers* has been reviewed elsewhere, their cost per transaction executed declined steadily over the past three decades (see Table 7). Thus, between 1954 and 1979, for a given number of dollars, processing speed increased by a factor of over 1,000. Put in nontechnical terms, the number of answers obtained from a computer for the same amount of money rose by over 1,000 times. The second most dramatic change came with computer memories. The price of disk storage declined by a factor of 100 between the mid–1950s and the late 1970s. Shifts in economics, as well as in the quantity and speed of technologies, were also evident. For example, in 1956 it was far cheaper to have data on magnetic tape than on disk and so batch applications dominated. But by the early 1980s the reverse was true, accounting for the sharp rise in online applications and direct access of data. Montgomery Phister notes that, when inflation was taken into account, the dynamics of dollars with storage were profound and informative. According to Phister, a penny in 1955 had the purchasing power of 2.5 cents in 1979. Using that baseline, he discovered that the cost of storage on disk per byte went from 2.5 cents in the mid–1950s to 0.006 cents, using 1979 cents. Thus, while in absolute numbers costs for storage had declined by a factor of 167 to 1, when recalculated to take inflation into account, it was really 417 to 1. A similar tale could be told for memories and processors. The only constant was card punch equipment which had essentially the same value throughout the entire period in question.

Although data on the characteristics of each decade are presented below, it is instructive to study the value of hardware by several groups over time. One of the earliest sets of users of computers were U.S. government agencies. As Table 8 suggests, during the short period from 1962 through 1974, the number of computers owned by the federal government grew fourfold. Taking into account raw dollars expended from 1954 through 1973 by all users for computers and

Table 8
Number of Computers Owned by the U.S. Government, by Management Classification, 1962–1975

Year	General Management and Business Applications	Special Management and Special Military	Total
1962	1,030		1,030
1963	1,326		1,326
1964	1,862		1,862
1965	2,412		2,412
1966	3,007		3,007
1967	2,754	938	3,692
1968	2,909	1,323	4,232
1969	3,039	1,627	4,666
1970	3,404	1,873	5,277
1971	3,389	2,545	5,934
1972	3,433	3,298	6,731
1973	3,432	3,717	7,149
1974	3,487	4,343	7,830
1975*	3,622	5,027	8,649

*Since the source was published in 1975, the total for that year may not be complete. These data leave out military systems installed before 1966. The U.S. military establishment was one of the earliest users of computers and dominated much of its development during the 1950s. As the data for 1967 suggest, hundreds had been acquired prior to that date by the U.S. military community.

SOURCE: General Services Administration, *Inventory of Automatic Data Processing Equipment in the United States Government*, Report No. 022001000663 (Washington, D.C.: GSA, 1975): 2.

related equipment in the United States, the growth in expenditures was one of the most significant trends within the industry. Table 9 presents a reliable set of statistics suggesting the magnitude of growth. The data processing industry grew, for example, by over 27 percent per year during the 1960s and frequently between 11 and 17 percent during the 1970s. It was slower than either of those two sets of statistics during the 1950s and 1980s.

Worldwide revenues generated by the sale of computer-related equipment between 1965—the year third-generation computers were sold in quantity—and 1981 also grew sharply. As Table 10 indicates, the second half of the 1960s witnessed a growth matched by no other major sector of the U.S. economy, and, while it slowed during the early 1970s—a period of economic recession in the United States—it nonetheless grew and experienced another takeoff in the late 1970s. Another way to measure the size of the industry is to determine the

Table 9
Expenditures for Computer Hardware in the United States, 1954–1973, Selected Years

Year	Millions Dollars
1954	10
1958	250
1963	1,500
1968	4,500
1973	7,000

SOURCE: *EDP Industry Report* (Waltham, Mass.: International Data Corporation, August 1974): 2.

Table 10
Worldwide Data Processing (DP) and Business Equipment (BE) Industry Revenues, 1965–1981, Selected Years (Billions of Dollars)

Description of Revenue	1965	1970	1975	1980	1981
Total	4.5	20.5	46.2	90.6	101.6
Total DP Equipment	2.4	10.5	27.9	55.1	61.8
Application of Software Excluding Contract Program	.2	2.5	6.5	18.0	21.0
Total BE Equipment	3.5	6.0	9.3	13.5	14.7
Business Form	.9	1.5	2.5	4.0	4.1

SOURCE: U.S. Department of Commerce, International Trade Administration, *The Computer Industry* (Washington, D.C.: U.S. Government Printing Office, April 1983): 52.

number of people employed by manufacturers of computers in the United States. Table 11, covering the short period of 1967–1972, when sales grew rapidly and then declined after the economic recession, showed a range of 145,000 to 190,000 jobs that did not exist a decade earlier. Tables 12 and 13 reconfirm the trends described earlier. Data processing enjoyed a similar, if smaller, experience in Europe during the same period.

In addition to the sale and service of hardware within the data processing industry, there were other sectors. One of these made up the service sector within the industry, and was known as service bureaus. These bureaus were companies or individuals who would either turn data into machine-readable form and produce

Table 11
U.S. Employment by Computer Equipment Manufacturers, 1967–1972, Selected Years

Date	Employment
1967 average	145,100
1968 average	160,600
1969 average	182,700
1970 average	190,300
1971 average	170,100
August 1972	172,300

SOURCE: U.S. Department of Labor as reproduced in Bruce Gilchrist and Richard E. Weber, eds., *The State of the Computer Industry in the United States* (Montvale, N.J.: AFIPS, 1973): 23.

reports for clients or make computing power available to their customers. In the early 1900s companies had been formed that owned or leased tabulating equipment used to punch data into cards and process them for clients with no such equipment. By the 1930s that kind of service had become an important subset of the office machine equipment environment. Some of these companies and others that were founded later bought some of the first commercially available stored-program digital computers to provide services to customers who knew nothing about data processing and did not own computers. By the end of 1976, just within the U.S. economy, there were over 1,500 such companies, perhaps the most important of which was Automatic Data Processing (ADP).†

At first, batch processing dominated, with, for example, weekly reports prepared and mailed to customers. In the 1960s telecommunications made it possible to transmit information back to a customer over a telephone line to a specialized terminal or printer. During the 1970s one could also access a computer via a terminal, an approach to computing that had first appeared at MIT in the 1950s and later in the 1960s across the entire U.S. economy. One simply paid for either "x" amount of time attached to a computer or a fixed rate for output. As Table 14 indicates, service industry sales grew from 12 percent of all nonhardware revenues in 1960 to over 20 percent by 1980. Measured in billions of dollars, revenues grew almost as rapidly as expenditures for hardware during the same period (see Table 15).

Until recently, software, though important, has been neglected by economists and historians. The large vendors in the industry did not focus on this market until the 1980s, even though they had led the entire industry in manufacturing programs for decades. But like economists and historians, the executives of these firms as a rule thought of software as ancillary to hardware. By the mid–1980s

Table 12
Growth Rates of Revenues of U.S. Firms from World Operations, 1961–1971

Period	General Purpose Computer Systems	Mini & Dedicated Application Computer Systems	Services	Leasing	Supplies
1961-1965	31.5	37.5	20.3	---	7.7
1965-1969	24.3	34.3	28.9	84.5	10.3
1969-1971	8.2	-6.0	20.5	6.4	3.9
1961-1971	23.0	27.5	23.8	58.5*	8.0

*For the period 1965–1971

SOURCE: International Data Corp., as reproduced in modified form in Bruce Gilchrist and Richard E. Weber, eds., *The State of the Computer Industry in the United States* (Montvale, N.J.: AFIPS, 1973): 21.

Table 13
Annual Exports and Imports of Computer Equipment in the U.S. Market, 1967–1971 (Millions of Dollars)

Exports-Imports	1967	1968	1969	1970	1971
Exports	475	530	786	1,237	1,262
Imports*	20	18	37	60	119
Net Exports	455	512	749	1,177	1,143

*Parts not included.

SOURCE: Bureau of Domestic Commerce, U.S. Department of Commerce, as reproduced in Bruce Gilchrist and Richard E. Weber, eds., *The State of the Computer Industry in the United States* (Montvale, N.J.: AFIPS, 1973): 21.

Table 14
Estimated Growth in Service Sector of Data Processing Industry, 1955–1979, Selected Years

Year	Percentage of Revenue (U.S. Market)
1955	17.0
1960	12.5
1965	11.0
1970	18.0
1975	18.5
1979	20.0

SOURCE: Montgomery Phister, Jr., "Computer Industry," in Anthony Ralston and Edwin D. Reilly, Jr., eds., *Encyclopedia of Computer Science and Engineering* (New York: Van Nostrand Reinhold, 1983): 342.

all of that was changing. The first to take software seriously were executives of manufacturing firms within the data processing industry who recognized that programming represented an untapped multibillion dollar opportunity. Understanding this segment of the industry proved difficult because historically 85 percent of all expenditures on software went toward salaries of employees to write programs. This situation stood in sharp contrast with hardware because users typically did not build their own computers; they leased or bought them

Table 15
Estimated Revenues from Service Sector of Data Processing Industry, 1955–1979, Selected Years (Billions of Dollars)

Year	Total
1955	0.3
1960	0.4-0.5
1965	0.5-0.7
1970	1.1
1975	3.2
1979	5.5

Facilities management (service sector) revenues remained almost miniscule until after 1970 when they grew, reaching nearly $500 million by the late 1970s; online services grew from several hundred million dollars in the late 1960s to over $3 billion by the end of the 1970s; batch processing experienced a less steeper growth rate than online processing but grew from approximately $400 million in 1965 to over $2 billion in 1980.

SOURCE: Montgomery Phister, Jr., "Computer Industry," in Anthony Ralston and Edwin D. Reilly, Jr., eds., *Encyclopedia of Computer Science and Engineering* (New York: Van Nostrand Reinhold, 1983): 342–343.

from manufacturing firms such as IBM or Honeywell. As of 1986 only some generalities concerning software can be presented.

In addition to the 85 percent spent on salaries for programmers and systems analysts, another 5 percent went to cover the costs of software manufacturing firms for developing programs. The remaining 10 percent was spent for renting or buying software packages or services. Most of the data available about expenditures on software concerned this 10 percent. That portion of software constituted the software industry or subset of the data processing community, and the greatest amount of data has become available for it over the past decade.

Vendors first emerged in the late 1940s and early 1950s, primarily as contractors servicing government-supported projects, usually concerning defense systems. Software was made to order during this period. By the mid–1950s, companies were developing application packages that organizations with specific types of computers could use. Programming languages,* for example, came into their own, along with accounting, financial, and inventory control programs. During the 1960s this trend continued, keeping up with new developments in hardware. In the early 1970s database management systems* flourished along with thousands of new application packages. By the early 1980s, for the IBM Personal Computer alone there were over 3,000 software packages on the market. As Table 16 strongly implies, the software community expanded into an important segment of the industry after the announcement of the IBM S/360* in

Table 16
Estimated Software Revenues, 1964–1980, Selected Years (Millions of Dollars)

Year	Custom Programming	Packages	Total
1964	150-175	50-100	175-275
1968	300	100	400
1970	300	200	500
1972	350	300-350	600-650
1974	400	300-350	700-750
1976	400-425	750	1,100
1978	500	1,000	1,500
1980	600	1,500	2,100

Based on available data, statistics for custom programming in the 1960s cited above may be 10 to 25 percent too low and for packages 10 to 15 percent too high; for the 1970s all data may be 15 to 25 percent too low. By the mid–1980s software amounted to $4 billion per year in the United States. All data above are for U.S. market only.

SOURCE: Extracted from figures in Montgomery Phister, Jr., "Computer Industry," in Anthony Ralston and Edwin D. Reilly, Jr., eds., *Encyclopedia of Computer Science and Engineering* (New York: Van Nostrand Reinhold, 1983): 343. Copyright © 1983. Used with special permission. All rights reserved. See also: James W. Cortada, *Strategic Data Processing: Considerations for Management* (Englewood Cliffs, N.J.: Prentice-Hall, 1984): 37–46.

1964. Revenues in the United States went from several hundred million dollars to over $2 billion by the late 1970s, and by the mid–1980s had doubled again.

The dynamics of software business has received very little attention in the literature. Several factors that influenced this segment are identifiable. First, in the 1950s there were too few people who knew about data processing and thus were forced to turn to specialized companies. Computer manufacturers such as IBM wrote all the systems control programs (SCPs) necessary to operate equipment and frequently the compilers necessary to use such major languages as COBOL,* FORTRAN,* BASIC,* and APL.* Industry groups also participated or defined the architecture for specific languages (such as for COBOL and ALGOL*). The number of lines of programming written in each decade is not known, but the statistics would probably be massive. Citing one small example to illustrate the issue, just for one SCP, IBM wrote over 5 million lines of code for the operating system of the S/360 in 1968, and that statistic did not include application software. Many minicomputers of the early 1970s had 500,000 to 700,000 lines of code. A major purchasing application package might also have a similar amount of programming. To put that size and quantity of code into greater perspective, in the 1970s it was estimated that one line of code cost $10 to write.

Another element was IBM's unbundling of software in 1969. In midyear, IBM announced that it would no longer "bundle" together software and hardware for one price. Rather, it would now break the two apart, with the exception of operating systems and other basic utilities. Hence, firms that sought to compete against IBM (as well as competitors who also unbundled) could do so by providing packages for applications, sort/merge utilities, and competitive database managers. Such a policy change probably did more to encourage the growth of a software industry than any other factor. Sales statistics for that segment suggest a correlation between unbundling and the growth of revenues. By the end of 1975 standard software packages exceeded customized programming in value.

Some issues of historical interest have influenced the use of software. During the 1950s and early 1960s, programming languages were complex and usable only by people with data processing backgrounds. As the number of computers installed increased, so too did the demand for programmers to write more application codes. By the early 1970s the entire industry was experiencing a shortage of qualified programmers who could deliver enough programs. By then, the use of data processing could mean life or death for many organizations. As the cost of computing dropped, demand for more application software rose, often far ahead of price improvements in hardware. With the introduction of microcomputers in the late 1970s, the need for more software that could be used and maintained by individuals with little or no background in data processing began reaching crisis proportions. The industry could not train enough programmers, and thus pressure increased to develop software faster, better, and with decreasing amounts of technical expertise by writers. At the same time, the industry searched for better ways to make existing programmers more productive—hence, the introduction of database managers, code generators, macros, and decision-support packages. Quite possibly then, once the history of programming is properly understood from the point of view of users of software, historians will have a firmer grasp of why and how the industry was so rapidly accepted beginning in about 1965.

In addition to (1) the dynamics of companies manufacturing hardware and their impact on the industry as a whole, together with (2) software vendors and programming, there is a third major component of the data processing industry: those who make and sell the consumables of the industry, primarily the media in which data are stored in machine-readable form. These products include computer cards, paper, disk packs, magnetic tape, and diskettes and, as a subset of the industry, existed from the day Hollerith decided to sell cards while leasing tabulators. Examination of the past thirty years points to some identifiable trends. First, the cost of computer paper rose sharply, especially since the 1960s. Second, computer cards remained essentially flat in terms of both price and quantity used since the 1970s. Third, a similar pattern existed with tape; the only exception came in the 1950s when this was the medium of choice over disk. Fourth, disk packs (hardly sold after 1980) made up the smallest portion of the media community and declined sharply in the 1980s. Diskettes (which look like 45 rpm phonograph records and are used primarily with personal computers) have not been around long enough to characterize generally. However, overall this medium has not been an inconsequential component of the data processing industry (see Figure 4). Total sales in the United States increased from several hundred million dollars in the 1960s to nearly $2 billion by the 1980s.

Price performance was evident with media as elsewhere in the industry. The 1970s witnessed the greatest improvement in price/performance since the birth of computers. Magnetic tape in 2,400-foot reels (the standard length), for example, went from $13 each in 1971 to just under $10 in 1979. Disk packs were widely used in that decade and declined in price from about $1,000 to $500 (for those used on the IBM 3330—the industry standard). The cost of paper, usually continuous forms and often with multiple copies and ribbons, often preprinted, soared by approximately 125 percent in the same decade. But that price increase affected all users of papers—newspapers, book publishers, and vendors of office supplies—as well as data processing. Finally, the cards that IBM sold in the 1930s for just over $1 per thousand dipped to just under $1 in 1971 but in 1978 reached $2.30 per thousand. Interestingly, IBM, which had so carefully guarded its ability to sell this product for nearly seven decades, discontinued their manufacture during the 1980s.

The only other major segment of the data processing industry was telecommunications, which involved transmissions and specialized equipment. As a subset, this segment came into its own in the 1970s and by the early 1980s attained a value of over $2 billion just within the United States. Companies that sold computers got into telecommunications (IBM, for example), while those strictly in telephony and telecommunications sold computers, American Telephone and Telegraph (AT&T),† for example. There was a great stirring in telecommunications, and its characteristics have been almost totally ignored by historians. Thus, it has become almost as difficult to understand telecommunications as the dynamics of software. Compounding the problem was the fact that TP, as it was known, did not become an important issue in data processing until the late 1970s.

Let us now examine the trends that characterized each period. The single most persuasive feature of the 1950s was the attempt to commercialize computers. The major companies of this period were Remington Rand which was the premiere company, particularly after acquiring Univac; later IBM became important, along with firms such as GE, Burroughs, and Honeywell. The first computer built in a factory (as opposed to within a user's data center) was the IBM 701,* which was announced in May 1952 and which signaled the company's commitment to participate extensively in the world of computers. The next major announcement was the IBM 650,* announced in 1953 and first shipped in 1954. That product propelled IBM into the mainstream of the new industry. Over 1,800 of these products were sold—a phenomenal number by the standards of the day.

Figure 4
Revenues for Media in U.S. Data Processing Industry, 1955–1979

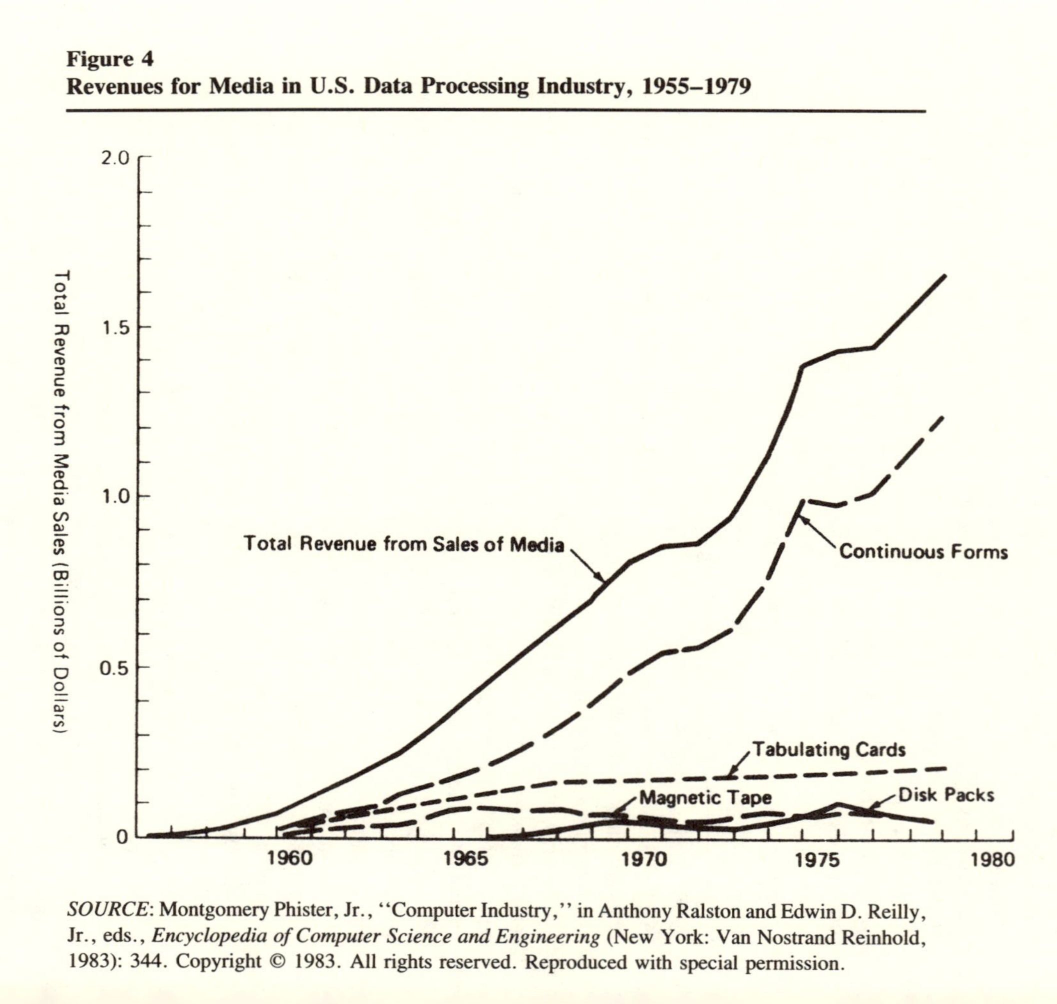

SOURCE: Montgomery Phister, Jr., "Computer Industry," in Anthony Ralston and Edwin D. Reilly, Jr., eds., *Encyclopedia of Computer Science and Engineering* (New York: Van Nostrand Reinhold, 1983): 344.

The 650 was likened to the Model-T of computing because it was the first mass-produced computer.

Such changes drove down costs and helped standardize the introduction of new technologies. By the end of 1954, the two dominant vendors were Remington Rand with its UNIVAC I (also brought out in the early 1950s) and IBM. In addition, twenty-three other companies that year had either manufactured computers or declared their intentions to do so. But only IBM and Remington Rand had made the necessary investments in developing products and marketing programs essential to success in what from the earliest days was a highly competitive market.

The decade was characterized by limited capacities in computers and equally limited knowledge of computing by potential customers. Everyone was a first-time user forced to invest a great deal of money in quasi-reliable technologies. There was real concern about quantifying the benefits of such expenditures. The confidence which would be so much in evidence a decade later was nowhere in sight. Vendors had to teach customers why they should use equipment and how to operate it, while trying to justify costs. At the same time, vendors rushed to improve reliability and capacity, particularly the speed and size of memories in computers. Major suppliers provided software, education, maintenance, and encouragement. Those tactics continued throughout the 1950s and 1960s.

The rivalry between IBM and Remington Rand was finally resolved by 1956 with IBM the clear winner. Along the way certain technical milestones were achieved which in turn spurred the use of data processing. SAGE,* a defense system sponsored by the government, generated a great deal of new technology which emerged by the late 1950s in various computers, especially those introduced by IBM. In the mid–1950s just over half of IBM's data processing revenues in the United States came from SAGE. The ability to use a high-level programming language called FORTRAN, beginning in 1957, began the long process of making programming easier. Before that date, almost all programming was done by using a machine-level code that required significant technical expertise by a programmer. With high-level languages, productivity increased because it took less knowledge to use, could be written quicker, and commands could be articulated with fewer instructions. Thus, for example, at General Motors, FORTRAN cut the time it took to write a program by a factor of five. By 1956 IBM had also made the critical decision to begin retiring its long-profitable tabulating business and had committed itself fully to computers. During the rest of the decade, it brought out a variety of computers and introduced the disk drive, making it possible to develop large online files and direct accessing of information during the 1960s.

Applications in the 1950s fell into two categories. At the beginning of the decade, *scientific users* dominated computing, particularly at universities and at government-managed research facilities. The government had its own applications covering a wide range of issues: the census, weather, atomic bombs, real-time computing for defense, inventory control, and payroll. By the second half of the decade, *commercial users* began to weigh in heavily as they sought to reduce clerical expenses and cut the number of employees doing tasks computers could perform. At first applications included traditional accounting (accounts receivable, accounts payable, general ledger, etc.), followed by payroll and inventory control. Analog computers* were used for process control in manufacturing. Almost everything was batch work with minor exceptions usually found at universities. MIT was one of the leading centers in the United States for innovations and home for some of the earliest uses of real-time systems and cathode ray terminals.

Remington Rand reflected these trends, although it did not succeed. The company began the decade in a better position than any other firm: it had the UNIVAC which was synonymous with the word *computer* during the first half of the decade. Unlike IBM, however, it did not commit itself totally to data processing and thus was not able to compete as effectively as needed in the late 1950s. Its UNIVACs came out late when newer technologies were driving costs and hence profits per unit down for all vendors. During the winter of 1954–1955 IBM gained the technical lead and kept it for the rest of the decade. Honeywell played a minor role, delivering its first computer in 1958 and selling only about ten such machines. GE built one machine and did not attempt to play an important role until the 1960s. NCR did not deliver computers between 1954 and 1959, and Radio Corporation of America (RCA)† produced only six. Most firms had failed to recognize the potential of computers early enough, and thus the industry as a whole remained small.

The advent of a second generation of computers in the late 1950s signaled increased attention and activity. IBM profited enormously from its STRETCH* project, which caused so much new technology to become available for use in products in the late 1950s and early 1960s. The SABRE* system, providing the first online reservation system in the world, was launched with American Airlines. That project gave IBM a wealth of technology which it applied in the years ahead. The airline set a standard with SABRE which the industry copied, and the application was still an integral part of that company's operations in the mid–1980s. The number of computers sold increased, suggesting that a new era was at hand while quantifying its size. IBM brought out the 1401* in October 1959 and sold almost 20,000 of them. To put that accomplishment in perspective, in the United States alone in 1960 only 6,000 general-purpose computers were installed from all vendors. In short, the data processing industry began its enormous and rapid growth at the dawn of the new decade. The period from the end of the 1950s to April 1964 was characterized as an historical era. It was the time of second-generation equipment, modest expansion in applications, and introduction of new technologies, most notably use of the transistor, with the chip* ready to appear. As Table 17 illustrates, the worldwide population of computers grew sharply by 1962, with the United States dominating the entire data processing industry.

By the early 1960s the story of data processing could no longer be told solely

Table 17
Number of Installed Computers, Worldwide, 1962

United States	9,337
Germany	472
Great Britain	389
Other Nations	761
Total	10,959

This list represents installed equipment at the height of second-generation technology and before the great spurt in the installation of computers caused in large part by the announcement of the IBM System/360 in April 1964.

through the rivalry between IBM and Sperry Rand Corporation† (the old Remington Rand) but now had to include other vendors entering the market. As measured in terms of data processing revenues only (citing 1963's data since it was at the highwater mark of second-generation gear), IBM was number one ($1.2 billion) but shared sales with, among others, Sperry Rand ($145.5 million), American Telephone and Telegraph (AT&T) ($97 million), Control Data Corporation (CDC)† ($84.6 million), Philco† ($73.9 million), Burroughs ($42.1 million), GE ($38.6 million), NCR ($30.7 million), and Honeywell ($27 million). IBM retained its dominance by concentrating on data processing markets, unlike other firms which were also interested in other products (such as RCA with television). It built its own in-house technical expertise and risked its corporate health on new technical ventures that did well. No other company was willing to invest proportionally as much on data processing, and so IBM prospered in the years before its announcement of the S/360, despite its own perception that the successful products of the past (e.g., 1401) were aging.

The most dramatic event of the 1960s was IBM's announcement in April 1964 of the S/360. It was to be the most successful product in the history of American industry. Under its blue covers it gathered the latest technology packaged in sensible ways that made use easier and at costs impressively lower than in the past. The results started the whole industry. By the end of the decade, IBM had doubled in size, and, overall, the industry grew each year at averages in excess of 27 percent. By 1968, 15 percent of all computers in the world were S/360s. This computer provided less conversion for users moving from one size machine to another, standard operating systems for all members of the computer's family which drove down operating costs, and improved reliability which encouraged people to "rely" on computers more than before. In effect, it ushered in the third generation of computers: all other vendors were now forced to scrap old products and bring out new, better, faster machines and in greater quantities.

Why was the S/360 necessary? A task force commissioned within the company produced the SPREAD report which argued that competition would eventually destroy IBM's ability to market what was already an aging set of computer products. To maintain and expand growth, IBM had to bring out better machines which took advantage of new technologies. The company's management agreed, poured every available resource it could get its hands on in the early 1960s into the project, and, in effect, built the giant that became the IBM of today. After the introduction of the S/360, every major vendor responded by lowering prices on their existing products while scrambling to introduce new ones that had many of the technical characteristics of IBM's new equipment. Furthermore, IBM became the target or benchmark against which great segments of the industry measured itself. The decade ended with IBM's announcement (followed by that of other vendors) that software would be unbundled.

Competition had heated up sufficiently in the decade to drive both GE and RCA out of the industry. In May 1970 GE announced that it had sold off its data processing business to Honeywell. In contrast to GE, which built computers different from IBM's, RCA manufactured computers (Spectra series) compatible with IBM's and marketed against the giant's installed base. In 1971 RCA made one of the most surprising announcements within the industry in withdrawing from data processing. Sperry Univac survived the 1960s and entered the 1970s as an important force within the industry.

During the late 1960s Burroughs converted fully into a data processing vendor. NCR introduced a few products, all of which were successful. CDC was one of the fastest growing companies between 1963 and 1969, with revenues increasing from $85 million to $570 million in that period. CDC prospered because it was practical and had many cost-effective products. Smaller firms with specialized knowledge were acquired and then integrated into the overall operations of a parent company, whose sole business was data processing. This pattern would be repeated many times within other organizations throughout the 1960s and 1970s.

The 1960s were characterized by features different from those evident in the 1950s. The first of these were manufacturers of plug-compatible equipment (called PCMs). Vendors, recognizing that IBM's equipment had set the standards for the industry in terms of technology, built computers and peripherals that used IBM's operating systems or that could be attached to configurations made up of its equipment. The plug-compatibles competed aggressively, primarily against IBM throughout the decade, first in the offices of potential customers and later with less success in the courtrooms of America. Some of the important PCM vendors were Telex,† Ampex, Memorex,† Information Storage Systems, and CalComp. They usually sold their look-alike products for between 10 and 15 percent below IBM's prices. They also had terms and conditions that gave them momentary advantages in one market or another.

Leasing companies were established during this period. They would buy a vendor's product, such as IBM's S/360s, and then lease them to customers over a longer period of time than would the manufacturer, thereby enabling reduced monthly costs for customers. This system also inhibited IBM from displacing such products as soon as it wanted with newer ones. Between 1961 and 1965 leasing companies bought between $10 and $20 million worth of IBM equipment. Cumulatively, they had acquired $2.5 billion worth of IBM products by 1969. Thirty-three percent of all S/360s purchased in 1966 were owned by leasing companies. That year there were 92 leasing companies; in 1970 the number had grown to 250 in the United States. Some of the major leasing firms of the period included Greyhound, Boothe Computer Corporation, and Itel.† Their stocks did well, until the 1970s when IBM and other major vendors brought out less expensive equipment to compete against purchased third-generation hardware.

Service bureaus did well in the 1960s. The Association of Data Processing Service Organizations proudly announced that, between 1965 and 1966, the revenues of the average firm grew by some 50 percent. In 1966 there were 700 service bureaus, and they generated $500 million in revenues. Some of the more important members of that community included McDonnell Automation, Automatic Data Processing (ADP),† and Tymshare.

Software firms were also thriving. By 1965 between forty and fifty independent software companies were operating in the United States, and they did 30 to 50 percent of all the systems software development work done in the United States for third-generation computers (many of them S/360s). Computer Sciences Corporation (CSC)† was particularly successful. By 1969, only ten years after its founding, its U.S. revenues reached $67.2 million. Informatics, Inc., established in 1962, ended the decade with revenues of $19.9 million. ADP, founded in 1959, also prospered and became an important supplier during the 1970s. In 1968 there were over 2,800 companies in this sector of the economy. Contract programming the following year approached $600 million; software products generated another $20 to $25 million in revenues just in the United States. Users spent $200 million writing their own software in 1960; $3 to $4 billion in 1965; and $8 billion in 1970. Five years later the volume had grown to $12 billion.

In the 1960s, then, data processing enjoyed a prosperity unrivaled by any industry in any period in American history. Only 6,000 computers had been installed in 1960, but by the end of 1968 the number had grown to over 67,000. The market had climbed to a value of $7.2 billion, reflecting a demand that continued to grow impressively during the 1970s. In 1972, despite a recession, total data processing revenues exceeded $6 billion. Of 618 firms in business in 1972, only 9 had existed in 1952, 75 in 1960, and 188 in 1964. Revenues in these years were $39.5 million, $1.3 billion, and $3.2 billion, respectively. The group in 1972 generated $12.8 billion. Although historians generally accept growth rates of over 27 percent, the above statistics suggest a compounded growth rate for the period that was closer to 33.5 percent.

The demand for computing set the pace, a requirement never saturated throughout the period. Colleges and universities, which historically have had less money to invest in leading-edge technologies, exhibited an impressive demand on manufacturers. In 1957 higher academic institutions had only forty-seven computers;

in 1959 that number jumped to 105 and the following year to 142. But then the takeoff came: 186 in 1961, 248 in 1962, and 329 in 1963. And IBM had not yet announced the S/360.

Demand could also be measured by the increase in the number of organizations using computers, creating a large base that continued to add data processing equipment to their inventories during the 1970s. The federal government, with only 5 computers in 1952, and 531 in 1960, 1,862 in 1964, and closed out 1970 with 5,277. In addition, as the 1970s wore on, newer machines were installed that had larger memories and greater capacities. All operated at faster speeds, thereby increasing the amount of horsepower available.

There are no reliable statistics on applications or usages in the private sector. But the previously cited statistics on the number of devices sold, the volume of dollars generated, and the improvement in price/performance ratios acknowledged were all dramatic, if indirect, descriptions of what happened during the 1960s in the growth of applications.

Historians have yet fully to assess the role of the federal government. As already mentioned, the government stimulated new technologies in the 1950s, primarily in an effort to miniaturize electronics for the military. After Sputnik in 1957, the Pentagon decided to invest heavily in rockets to deliver firepower. The addition of the U.S. space program and the creation of the National Aeronautics and Space Administration (NASA) in 1958 created yet more demand for miniaturization of electronics and for greater reliability and intelligence in such components. The electronics industry responded with new developments that were as dramatic as the computer chip, as mundane as better wiring of equipment, and as sophisticated as computer programming and even the use of new materials. The government's role attained giant proportions by the early 1960s. In 1961 it sponsored 70 percent of all basic research in every field of scientific endeavor in the United States. It never lost faith in that investment, and it poured much larger sums into research throughout the 1960s. On a more day-to-day level, the General Services Administration (GSA) actively encouraged the use of computers to improve routine governmental operations.

With regard to the size of the industry in the 1970s, data from 1972 are most suggestive. Some 5,000 firms supplied products to the data processing industry, nearly 700 of which made hardware and 500 of which made supplies of various types. The others produced a wide gamut of services. In 1971 the total value of hardware installed was $30.9 billion. In that same year there were 54,470 general-purpose systems in existence, of which only a third were very large (that is, leased for over $40,000/month). Yet this top third accounted for 75 percent of the total value of the installed base of computers. Minicomputers accounted for nearly 30 percent of all installed computers.

Shipments to customers outside of the United States indicate that, during the 1970s, other countries experienced what the United States had in the 1960s. In 1971 nearly 45 percent of all U.S. made computers went outside of the United States for a total value of $3.3 billion. As of December 1971 approximately

57,800 systems of all types and sizes were installed outside the United States (communist countries were excluded from this census), with a value of $18.4 billion.

Another way of measuring the size of the industry at the start of the decade involves employment. One study sponsored by AFIPS placed the total population in August 1972 at 172,300 but only included those employed in the actual manufacture of computers. Thus, the total industry (including all data processing professionals) probably was several times larger, perhaps closer to 300,000 that year. In 1971, American users of data processing expended about $8.3 billion on salaries, another $10.3 billion on goods and services, and probably another $5 billion on miscellaneous overhead, according to AFIPS. Put another way, in 1971 the data processing industry may have had a $23.6 billion or 2 percent share of the gross national product (GNP).

The decade began with some 24,000 organizations, each operating one or more computers; perhaps an equal number used service bureaus. Measured in terms of value and per $100 of national income in 1971, banking devoted 6.5 percent of its workforce to data processing. Ironically, railroads, which had been such a major user of tabulating equipment in the early years of the twentieth century, now only spent 0.7 percent of their labor on data processing in 1971. Education dominated per capita usage of data processing, followed in descending order by banking and insurance. If measured by percentage of employees devoted to data processing, the breakdown was: banking (6.4%), insurance (3.5%), aerospace and defense (2.3%), utilities (2.2%), airlines (1.5%), electronics (1.4%), and chemicals (1.2%), while each of the other sectors dropped below 1 percent. The total number of data processing professionals in 1971 had already reached 1 million in the United States—a useful yardstick by which to measure the growth of the industry even in the late 1960s.

The role of data processing outside of the United States increased sharply during the 1970s. As Table 18 illustrates, by examining the percentage of GNP devoted to data processing, one quickly realizes that the industry had become important to the United States (about 2.59 percent) and United Kingdom (2.13 percent), and was growing in significance to Denmark (1.71 percent), Switzerland (1.63 percent), and Canada (1.48 percent). Japan followed closely behind with 1.44 percent of GNP. The industry in the Soviet Union hovered at 0.40 percent, which suggests that it had not yet realized the significance of data processing to the local economy. Percentages of GNP should be used with caution because the definition of an industry often influences the statistical outcome. The broader the definition, the higher the GNP that would be cited. Because various government agencies and private researchers gather together such data, conflicting statistics on the size of the industry are common. However, the numbers suggest a relative presence within a particular national economy.

As early as 1969, the United States had 57 percent of all computer installations in the world compared to Japan and Germany which had 8 percent each. By the end of 1972 Japan had the second highest number of computer installations in

Table 18
Value of Installed Computers Compared to Gross National Product, 1973

Country	Installed ($ millions)	1972 GNP ($ billions)	Data Processing (% of GNP)
Japan	4,922	341	1.44
West Germany	3,772	292	1.29
United Kingdom	3,231	152	2.13
France	3,012	221	1.36
USSR	2,195	549	0.40
Canada	1,530	103	1.48
Italy	1,329	118	1.13
Australia	643	52.7	1.22
Netherlands	642	50.0	1.28
Switzerland	579	35.6	1.63
Sweden	459	43.7	1.05
Denmark	388	22.7	1.71
Belgium	338	38.8	0.87
Brazil	330	49.8	0.66
Spain	324	50.7	0.64
South Africa	233	19.1	1.22
Austria	181	22.8	0.79
Mexico	150	39.6	0.38
Norway	143	16.2	0.88
Yugoslavia	135	23.7	0.57
All other international	1,475	942	0.16
Total international	26,011	3,184	0.32
Total United States	29,942	1,155	2.59
Total Worldwide	55,953	4,339	1.29

Based on research conducted by the Bureau of Intelligence and Research, U.S. Department of State in 1974, showing that the United States led the world in the acquisition of computerized technology.

SOURCE: U.S. Department of State, *Bureau of Intelligence and Research Report* (Washington, D.C.: U.S. Government Printing Office, 1974): passim.

the world. Japan had around 45,000 general-purpose computers in 1970, 75,000 five years later, and 125,000 in 1980. The number of special-purpose computers increased from 25,000 in 1970 to 575,000 in 1980. In 1975 the United States had 8,649 machines, up 350 percent since 1965. The Department of Defense used 5,027 of these systems, making it the largest single user of computers within the government. Government use of computers grew steadily through the remaining years of the decade.

The 1970s witnessed technological innovations which led to continued decreases in cost and improved reliability. The major advances were use of monolithic semiconductors with large-scale integration, major advances in telecommunications, wide use of time-sharing services, and installation of minicomputers, largely for scientific and engineering computing or for distributed processing. High-speed printers, large storage and memory units, many new programming languages, and more usable operating systems rich in function also characterized the period. By 1979 Americans were living in the computer age. *Future Shock* was not simply the title of a book forecasting the future and increased use of computers. Americans sensed that they were experiencing its predictions daily. Almost all of their bills and paychecks were generated by computers, and credit checking had become a reality for over half the entire population.

Like other vendors, IBM replaced its third-generation equipment with fourth-generation computers. These were the S/370 and, in 1979, the 4300 series. IBM had already started to replace large S/370s with 308X machines. Computer wars now characterized the entire industry. Much of the fighting took place in courtrooms, with most lawsuits directed against IBM in an attempt to charge it with monopolistic practices. The list of litigants read like a *Who's Who* of the industry: Greyhound filed against IBM in 1969, Telex in 1972, CalComp, Hudson, Marshall, Memorex, and Transamerica in 1973, Forro and Memory Tech in 1974, and Sanders in 1975. They all lost. But the giant lawsuit was *U.S.* v. *IBM*, which was filed in January 1969 and lasted until the government dropped it in January 1982. It was the largest and longest antitrust suit in American history. The government failed to build a case against IBM. The case was important, not just as a legal precedent, but also because it influenced the actions of vendors throughout the decade. Its one great benefit was that it made public several billion documents on the entire data processing industry from sources scattered across the entire industry.

Although lawsuits acquired their own life-forms, major vendors competed in the marketplace. These included Burroughs, Control Data, NCR, Sperry Rand, Digital, Honeywell, and even Xerox† for a while. PCMs were very active, again targeting their efforts at IBM's S/360 and S/370 customers. Telex, Storage Technology, Memorex, and Advanced Memory Systems were active throughout the decade. As IBM introduced new computers or disk drives, for example, these vendors brought out similar products but at lower prices within eighteen months.

Leasing companies gained importance in the 1960s but faced a difficult period in the 1970s, particularly during the second half of the decade. They placed their equipment on depreciation schedules that were longer than IBM's, usually seven to ten years, based on their conviction that IBM would not announce replacement products before then. To their surprise, however, IBM brought out new products faster than it had in the 1960s, which reduced the residual values of leasing companies' machines faster than they were prepared to handle. They had no choice, therefore, but to depreciate faster, taking losses in order to finance inventories.

Another feature of the 1970s was the introduction of widespread use of minicomputers. Although they had been around since the early 1960s, not until the 1970s did they come into their own as departmental systems or as small, high-function application machines widely marketed by such firms as Data General Corporation,† Hewlett-Packard,† Prime Computer,† Perkin-Elmer, Harris Corporation,† Wang Laboratories, Tandem, and Datapoint.† These vendors did so well during the 1970s that the industry could no longer be discussed simply in terms of large mainframe manufacturers.

Plug-compatible large mainframes added competitive spice, particularly in 1975, when Amdahl Corporation† introduced its first product to sell directly against IBM's S/370s, particularly the larger models. They all did well during the mid–1970s, but with new introductions from IBM in the last years of the decade they experienced slowdowns in sales.

Software firms continued to thrive. In 1975 about 1,000 such companies with over 3,000 products in their sales kits had been established. When combined with contract programming, their revenues in 1975 amounted to $1.3 billion. Service bureaus also grew; ADP (twenty-one years old in 1970) expanded into one of the most important of these. By 1979 it had over 75,000 users generating revenues of $371 million.

IBM continued to be an important force in the market. According to data submitted in the government's lawsuit against the giant, IBM's share of data processing revenues in 1973 was 33 percent of all money raised in the industry, down from 49 percent in 1963. That company's domestic revenues during 1976 reached 44 percent of the total but by the end of 1979 had settled down to 34 percent. The market expanded, but IBM did not grow as quickly, which led management to launch an aggressive program introducing new products and spending over $17 billion in modernizing plants between the late 1970s and early 1980s. The results were spectacular during the early 1980s, but they were not evident in the late 1970s. Nonetheless, the company continued to prosper, though not with the growth rates it had experienced during the 1960s. The outstanding news for the company was the settlement of almost all private lawsuits against it, including the one filed by the Department of Justice.

An enormous variety of competition now typified the industry. Thousands of vendors which had not been in business at the start of the 1970s, let alone in the 1960s, were now operating in the industry. The use of data processing expanded to the point where the average company spent over 2 percent of its budget on it. Nor did technological change slow down. It was during this period, for example, that the desktop computer was born (late 1970s) which so profoundly influenced access to computers in the 1980s. It did nothing less than democratize computers.

During the 1970s it became easier to use computers, terminals, and even instructions that arrived with new equipment and software. Getting computers to do one's bidding by wiring boards in the 1950s or writing in complex languages in the 1960s was nothing compared to the ease that came with terminals in the 1970s using English-like instructions, a trend that continued into the 1980s. Terminals came into their own as the most widely used method for getting to computing power. Over 200 vendors sold these, primarily cathode ray terminals (CRTs). In the United States alone, *Computerworld* reported that, at the end of 1976, over 58,000 CRTs were installed along with 294,500 other terminals. The first 100,000 terminals sold in the United States went into data centers, but soon after they began appearing on desks in end-user departments, used by people who knew little or nothing about data processing. Whereas at the start of the decade the ratio of terminals to workers exceeded 1:100, by the end of the decade it was less than 1:10 and by the mid–1980s, 1:4, and was headed rapidly toward closer ratios.

In 1978 revenues of U.S. companies in the data processing industry accounted for approximately 1.5 percent of total GNP. It took the industry twenty-seven years to reach that point. To put that growth into perspective, factory sales of automobiles reached that percentage of GNP in about fifteen years and the telephone in forty-four years. Television sales reached 0.05 percent in less than a decade and remained in that constant level. Automobile sales varied between 1 and 3 percent of GNP and telephones at about 2.3 percent of GNP, while data processing (or information handling, depending on whose statistics and definitions were used) exceeded 1 percent by the end of the decade.

The volume of revenues generated by the industry made it a critical component of the U.S. economy by the early 1980s. As in the 1970s dominance of world sales by U.S. firms ranged between 70 and 80 percent, generating a trade surplus of $6.84 billion in 1982, despite challenges from Japanese vendors. The entire market was valued at $65 billion worldwide in 1982. Growth in the industryduring the 1970s averaged 18.1 percent and closer to 21.3 percent during the first three years of the 1980s. In 1976 worldwide production of U.S. computer companies was worth $23.4 billion but in 1981 had grown to $56 billion. The fastest growth came in software and services, which went from $5.2 billion to $14.9 billion. Mainframes nearly tripled in worth, from $6.2 billion to $17.2 billion, whereas minicomputers quadrupled to $8.8 billion. Only peripherals experienced slow growth, going from $10 billion to $13.9 billion.

The rate of growth in production by country also suggests the size and source of goods and services. Between 1978 and 1981, U.S. production grew by 23.2 percent, yielding a world market share of 57.7 percent; in second place was

Table 19
Sales of Major Data Processing Firms in Japan, 1976–1982, Selected Years (Millions of Dollars)

Company	1976	1979	1981	Growth Rate 1976–1981 (%)	1982 (Estimated)	Change Over 1981
Fujitsu	1086.5	1481.9	2033.3	13.4	2426.0	19.3
NEC	516.9	910.1	1507.7	23.9	1766.2	17.2
Hitachi	643.9	979.5	1305.9	15.2	1496.4	14.6
Oki	219.0	284.8	494.7	17.7	594.0	20.1
Toshiba	268.4	228.5	430.8	9.9	521.5	21.1
Mitsubishi	145.1	240.3	331.0	17.9	399.0	20.5
Subtotal	2879.8	4125.1	6103.4	16.2	7203.1	18.0
IBM Japan	1248.8	1470.1	1944.9	9.3	N/A	N/A
Nippon Univac	320.1	333.7	412.2	5.2	453.5	10.2
Total U.S. Affiliates	1568.9	1803.8	2357.1	8.5	N/A	N/A

SOURCE: *Japan Economic Journal* (June 9, 1981, June 8, 1982), Bureau of Industrial Economics, as tabulated and presented in U.S. Department of Commerce, *The Computer Industry* (Washington, D.C.: U.S. Government Printing Office, 1983): 25.

Table 20
Number of Microcomputers Shipped to Businesses, 1981–1985

Year	Volume
1981	344,000
1982	926,000
1983	1,538,000
1984	2,384,000
1985*	3,290,000

*Estimated.

SOURCE: Research was conducted by the Dunn & Bradstreet Corporation and appeared in *USA Today*, June 16, 1985, p. 5.

Japan with rates of 17.5 and 13.1 percent, respectively; and in third place was France with 18.1 percent and 9.5 percent, respectively. All other nations became less important except for Italy which grew 30.4 percent, apparently suggesting that it was catching up with other industrial countries, particularly its peers in the European Common Market. Italy's world share was only 2.3 percent, proof that most of its production was for internal consumption. When comparing worldwide manufacturing and buying patterns from one decade to another, during the 1980s the export of Japanese products was a new feature of the industry. Between 1980 and 1981 alone, Japan's data processing exports rose an average of 78 percent to nearly $700 million in value. Table 19 presents some of the characteristics of the Japanese data processing industry.

Unlike earlier decades, the 1980s made a wide number of microcomputers available. These small, portable computers could easily be placed on a desk at the office or at home. They first appeared in the late 1970s and in a few years had become a major product. The growth in the volume of micros and patterns of usage are shown in Tables 20, 21, and 22. The democratization of computing became evident as these little devices appeared in classrooms, college dormitories, and church rectories. Housewives kept recipes on them, children played games with their PCs, while uses in business and science made the traditional terminal an antique. Sales from 1978 to 1983 jumped from $15 million to nearly $500 million (Tables 23 and 24).

Computers could be placed in the home at a cost of less than $2,000 in the early 1980s, which would have cost over $1 million in the late 1960s. Decision-making increasingly relied on the modeling capabilities of such machines and their associated, relatively easy-to-use software. Social commentators began arguing that the way people thought and talked was being restructured to conform to the systematized approach of data processing.

During the past thirty-five years, data processing has become as visible and

Table 21
Size of Companies versus Percent Using PCs, 1985

Number of Employees	Percentage of Companies
1-19	23.9
20-99	36.3
100-499	47.2
500-999	71.8
1,000 or more	85.4

SOURCE: Research was conducted by the Dunn & Bradstreet Corporation and appeared in *USA Today*, June 16, 1985, p. 5.

Table 22
Business Uses of PCs, 1983 and 1985

Application	% in 1983	% in 1985
Accounting	61.3	72.5
Financial analysis	68.7	65.2
Word processing	48.4	56.8
Database management	34.6	38.3
Inventory control	30.5	31.5
Purchasing	14.1	22.8
Credit analysis of customers	20.3	14.2
Other	8.5	78.0

The data above came from a survey conducted by the Dunn & Bradstreet Corporation. Respondents were allowed to select more than one category; all were in the United States. Note the large variety of new applications that were obviously in use in 1985 as listed in the category called "Other."

SOURCE: *USA Today*, June 16, 1985, p. 5.

probably as important to society as the automobile, television, radio, or the telephone—and in a much shorter period of time. As Table 25 illustrates, the volume of devices installed was enormous.

Data processing flourished because of several factors. First, technological evolution in the physical sciences and in electronics came together in the mid-twentieth century, making it feasible to produce useful computational devices quickly. Second, the rate of change in that technology forced the providers of such equipment and software to compete, creating even more force for change and improvements. One could argue that competition within data processing

Table 23
Personal Computer Sales in the United States, 1978–1983 (Millions of Dollars)

1978	15
1979	30
1980	85
1982	180
1983	300-500

Data for 1981 was not provided in the original charts.

SOURCE: A. D. Little, Inc., published in Ulric Weil, *Information Systems in the 80's: Products, Markets, and Vendors* (Englewood Cliffs, N.J.: Prentice-Hall, 1982): 214.

Table 24
Personal Computer Systems in the United States, 1978–1983 (Thousands of Units)

Year	Annual	Installed
1978	20	30
1979	20	50
1980	70	120
1982	120	240
1983	600-1,000	1,600-2,000

Data for 1981 was not provided in the original charts.

SOURCE: A. D. Little, Inc., published in Ulric Weil, *Information Systems in the 80's: Products, Markets, and Vendors* (Englewood Cliffs, N.J.: Prentice-Hall, 1982): 214.

worldwide was perhaps as intense as that in any industry. Indeed, some have argued that nowhere was competition greater than within data processing where the technological imperative toward change and better performance pushed new companies into the business and kept the existing ones sharp or dead in their tracks. Third, convenience led to acceptance of data processing just as it had for tabulating equipment. People used things that were useful, and the power of computers to manage data was too irresistible to ignore. Throughout history humankind has quickly adopted the tools invented to manage information: writing, paper, printing, telephones, television, radio, and now computers.

A great deal of information is known about the number of machines built and shipped, their cost, and the worth of the industry, as is suggested by Table 26. We can also identify the major vendors and describe their histories. What has not been fully studied are the users and the uses of such technologies.

We also know little about the history of data centers, which represented homes

Table 25
Annual Shipments of U.S. Firms and Year-end Installed Base within the United States, 1966–1971

	General Purpose Computer Systems			Mini and Dedicated Application Computer Systems		
	Value of Shipments ($ millions)	Installed Base		Value of Shipments ($ millions)	Installed Base	
Year		Number (thousands)	Value ($ millions)		Number (thousands)	Value ($ millions)
1966	2,760	30.0	10,200	75	4.0	565
1967	3,815	35.6	13,300	125	6.0	690
1968	4,650	41.4	17,000	180	9.5	865
1969	4,642	46.5	20,800	277	16.1	1,135
1970	3,948	48.4	24,100	307	25.0	1,425
1971	4,074	54.4	26,400	249	33.5	1,665

SOURCE: "EDP Industry Report," March 1972, International Data Corporation, as reproduced in modified form in Bruce Gilchrist and Richard E. Weber, eds., *The State of the Computer Industry in the United States* (Montvale, N.J.: AFIPS, 1973): 11.

Table 26
World Computer Production by Country, 1978–1981

Country	1981 Output Value ($ billions)	Growth Rate (1978-1981)	World Market Share
United States	29.53	23.2%	57.7%
Japan	6.70	17.5%	13.1%
France	4.88	18.1%	9.5%
West Germany	3.50	13.3%	6.8%
Great Britain	2.33	12.2%	4.6%
Italy	1.19	30.4%	2.3%
Others	3.07	———	6.0%
Totals	51.20	20.6%	100.0%

SOURCE: U.S. Department of Commerce, International Trade Administration, *The Computer Industry* (Washington, D.C.: U.S. Government Printing Office, April 1983): 19.

for most computers until the early 1980s. We have only a few details about how such technology was purchased or leased, who made these decisions or precisely why. Nowhere is the ignorance greater than in the area of programmers and systems analysts—the hundreds of thousands of people who write the programs and design the applications. Finally, with the advent of microcomputers, we have to define who is using them and for what applications. Details on these issues and others will flesh out an otherwise purely statistical and economic description of what comprised the data processing industry.

For further information, see: James R. Beniger, *The Control Revolution: Technological and Economic Origins of the Information Society* (Cambridge, Mass: Harvard University Press, 1986); James W. Cortada, *An Annotated Bibliography on the History of Data Processing* (Westport, Conn.: Greenwood Press, 1983) and *Strategic Data Processing: Considerations for Management* (Englewood Cliffs, N.J.: Prentice-Hall, 1984); John Diebold, ed., *The World of the Computer* (New York: Random House, 1973); Franklin M. Fisher et al., *Folded, Spindled, and Mutilated: Economic Analysis and U.S. v. IBM* (Cambridge, Mass.: MIT Press, 1983) and *IBM and the U.S. Data Processing Industry: An Economic History* (New York: Praeger Publisher, 1983); Katharine D. Fishman, *The Computer Establishment* (New York: Harper & Row, 1981); Bruce Gilchrist and Richard E. Weber, eds., *The State of the Computer Industry in the United States: Data for 1971 and Projections for 1976* (Montvale, N.J.: AFIPS, 1973); Charles P. Lecht, *The Waves of Change: A Techno-Economic Analysis of the Data Processing Industry* (New York: Advanced Computer Techniques Corporation, 1977); Edward J. Lias, "A History of General Purpose Computer Uses in the United States, 1954 to 1977 and Likely Future Trends" (Ph.D. diss., New York University, 1979); Montgomery Phister, Jr., *Data Processing Technology and Economics* (Santa Monica, Calif.: Santa Monica Publishing Co., 1976); Anthony Ralston and Edwin D. Reilly, Jr., eds., *Encyclopedia of Computer Science and Engineering* (New York: Van Nostrand Reinhold, 1983); Everett M. Rogers and Judith K. Larsen, *Silicon Valley Fever: Growth of High-Technology Culture* (New York: Basic Books, 1984); Robert Sobel, *IBM: Colossus in Transition* (New York: Times Books, 1981); U.S. Department of Commerce, *The Computer Industry* (Washington, D.C.: U.S. Government Printing Office, 1983).

ORGANIZATIONS

A

ACM. See ASSOCIATION FOR COMPUTING MACHINERY

ADP. See AUTOMATIC DATA PROCESSING

AEDS. See ASSOCIATION FOR EDUCATIONAL DATA SYSTEMS

AFIPS. See AMERICAN FEDERATION OF INFORMATION PROCESSING SOCIETIES

AMDAHL CORPORATION. Amdahl, one of the more successful producers of plug-compatible computers, competed against International Business Machines Corporation's (IBM's)† large processors in the 1970s and early 1980s. Its history mirrors that of other vendors who sold machines that were similar to IBM's but that sold for less. The company's history was also punctuated by the founder's bitter rivalry with IBM.

Amdahl Corporation was founded in 1970, the product and brainchild of Gene Amdahl,** an IBM employee for nearly fourteen years and one of the chief architects of the S/360* family of computers. In the late 1960s he sought to convince IBM of the need to produce yet more advanced systems and larger processors, but, for marketing reasons, IBM chose not to at that time. Amdahl left the firm and established his own. His idea was simple: to develop and build large computers for general-purpose applications that would be compatible with IBM's, and yet would cost less. To help finance his project, he shared ownership with a Japanese computer manufacturer, Fujitsu, Ltd., which, in 1975, had 41 percent of the company's stock. Heizer Corporation, a venture capital firm, had 31 percent interest. Amdahl's company was one of the most successful of the plug-compatible corporations. In part, Amdahl attributed his success to useful products, but he also kept his costs down by having Fujitsu build subassemblies for computers.

Several years elapsed before Amdahl began introducing products. In the middle of 1975 his company brought out the 470 V/6 (pronounced V as in Victor six) computer that could serve as a one-for-one replacement of IBM's largest S/370* computer, the Model 168. This and subsequent products used IBM's operating systems* and peripheral equipment, thereby keeping development costs down. The V/6, along with other models generated revenues for Amdahl which, in 1978, reached $321 million. Some of the key products contributing to the company's early successes included the 470 V/6 II processor aimed at the IBM 3032; the 470 V/7 comparable to the IBM 3033; the 470 V/5 for users of the IBM 3031 or the 370/168–3; and the 470 V/5-II and the 470 V/8 in the same size range as the IBM 3033. In 1978 Amdahl Corporation also introduced software products to optimize its equipment when used with IBM's operating systems. The most important of these products included MVS/SE Assist and the VM/Performance Enhancement, all of which became available during the early 1980s.

Amdahl Corporation was one of many plug-compatible manufacturers (also known as PCMs) which in the 1970s made products that competed for every component of an IBM hardware configuration. They all decided not to design operating systems inasmuch as they would be different from IBM's and would therefore force customers to convert their programs, something none of them wanted to do. Furthermore, the cost of developing such software would have exceeded that for hardware. PCMs included National Semiconductor, Fujitsu, Hitachi, and Magnuson (which sold against IBM's 4300 and employed Gene Amdahl's son, Carlton). In 1979 sales of plug-compatible devices declined for all major vendors, including Amdahl, in anticipation of a new generation of IBM equipment. Indeed, the large computer company announced the 4300 family of computers (4341s and 4331s in February, with schedule dates in June and first shipments in the fall) and other members of its large-end processors, representing new levels of price/performance improvements using newer technologies. Following a lull in business, Amdahl sold stock to raise cash. As a result, the founder of the company, Gene Amdahl, relinquished control of his firm to Fujitsu, which emerged with 49 percent ownership. Additional investments in the company by Fujitsu during the early 1980s revived the company, and in 1983 Amdahl generated $778 million in revenues and earned $46 million on these sales. In 1984 revenues flattened out at $779.4 million. By then the company had a workforce of 7,000 employees building and selling in Canada and in the United States. *Datamation* ranked the company thirty-sixth out of a 100 in terms of gross sales within the industry.

Throughout the late 1970s and early 1980s, Amdahl remained the largest plug-compatible vendor in the data processing industry. In 1984, after expensive delays, the corporation shipped a new family of processors which were reasonably compatible with IBM's current operating system, MVS/XA. In 1985 and in

1986, 6380 disk drives were marketed against IBM's 3380 disk drives, promising to contribute almost as much in revenues as processors.

Because of Gene Amdahl's connection with IBM, many industry pundits viewed his company's efforts as evidence of a personal rivalry between himself and his old employer. As noted earlier, Amdahl lost control of his firm in 1979 when Fujitsu increased its ownership of stock in the company, which now began using Amdahl Corporation as a channel for its technologies and products (including disk, computers, and other peripherals) into the United States. In 1980 Amdahl started another company, Trilogy Systems Corporation, which represented one of the largest startup operations in the data processing industry. Since 1980 nearly $300 million has been invested in the firm to build faster, less expensive computers using advanced superchips. Investments came from many sources in large part spurred by confidence in Amdahl: Sperry Rand Corporation* invested $42 million, Digital Equipment Corporation (DEC)† $26 million, Honeywell† $13 million, and Control Data Corporation† $2 million, all in exchange for being allowed to use his wafer-scale technology if it was ever perfected. Amdahl ran into difficulties in delivering the new technology, and as of 1986 had not shipped major products. In the fall of 1985 he turned his attention to his third and newest venture, a company called Elxsi which sought to build computers in a market for a machine larger than a minicomputer but no larger than a supercomputer.

For further information, see: Franklin M. Fisher et al., *IBM and the U.S. Data Processing Industry: An Economic History* (New York: Praeger Publishers, 1983); Thomas O'Donnell, "Gene Amdahl's White Whale," *Forbes*, June 18, 1984, pp. 46, 50.

AMERICAN FEDERATION OF INFORMATION PROCESSING SOCIETIES (AFIPS). Historically, AFIPS has represented the data processing industry as a whole while sponsoring numerous educational programs. It is made up of other associations within the data processing industry and, as of 1986, consisted of eleven associations.

AFIPS was established on May 10, 1961, through the efforts of the National Joint Computer Committee (formed in 1951) whose chief purpose was to organize annual Joint Computer Conferences. This federation of societies was created to proliferate and diffuse knowledge about information processing science in general. Much like a lobby almost from its inception, it always provided considerable information about industrywide issues. Some recent concerns include legislative issues before the U.S. Congress, data security, and education. Two of its primary programs are the annual Joint Computer Conferences held in the spring and fall. In 1973 these conferences merged to form one annual event called the National Computer Conference.† In addition, AFIPS has sent representatives to various international meetings.

When it was established in 1961, associated members were the American Institute of Electrical Engineers and the Institute of Radio Engineers (today the two are the IEEE†), and the Association of Computing Machinery (ACM).† By 1975 membership had grown to fifteen: ACM, IEEE, Data Processing Management Association (DPMA),† Society for Computer Simulation (SCS—at one time Simulation Councils, Inc.),† Association for Computational Linguistics (ACL),† Society for Information Display (SID),† Special Libraries Association (SLA), American Statistical Association (ASA), Society for Industrial and Applied Mathematics (SIAM), American Institute of Aeronautics and Astronautics (AIAA), Instrument Society of America (ISA),† Association for Educational Data Systems (AEDS),† the Institute of Internal Auditors (IIA), American Society for Information Science (ASIS),† and the American Institute of Certified Public Accountants (AICPA). By 1985 reorganizations and consolidations, in large part by its member societies, had reduced the group to eleven: ASIS, ASA, ACL, ACM, AEDS, DPMA, IEEE, ISA, SCS, SIAM, and SID.

AFIPS's educational programs have been important events. Biannual gatherings were held until 1973 when they were replaced by the National Computer Conference. These events have involved thousands of visitors, hundreds of vendors displaying their products, and dozens of seminars on industry issues and on ways to better manage data processing. These activities have reflected all facets of the industry and academic pursuits. In addition, AFIPS's publications list includes dozens of book-length studies, proceedings of annual conferences by member associations from the early 1960s to the present, and specialized monographs concerned primarily with technology and the management of data processing.

AFIPS has been an important voice for the data processing industry as a whole. In addition to representing American data processing organizations in international conferences, its members have also supplied witnesses before U.S. congressional hearings on issues relevant to the data processing (DP) community as a whole. Its publications have helped to disseminate information about data processing technology through books, journals, proceedings, and a marketing organization distributing publications issued by member organizations. It has also sponsored efforts to coordinate activities and educational programs of member societies.

By the mid–1980s the total members in AFIPS's constituent societies totaled over 240,000. As such, AFIPS spoke for an important community within American industry. It represented the American data processing community as delegate to the International Federation for Information Processing (IFIP†)—yet another association but one whose members are drawn from societies around the world. IFIP was established in 1959 under the auspices of UNESCO in recognition of the growing importance of data processing around the world.

AFIPS recognizes outstanding work by individuals within the industry through its Harry Goode Memorial Award, initiated in 1964. Some of the earlier recipients included Howard H. Aiken**(1964), a pioneer researcher working with digital computers,* George R. Stibitz**(1965), who developed relay computers at Bell

Laboratories,† and Konrad Zuse**(1965), a pioneer in the manufacture of German computers in the 1940s. In 1966 AFIPS honored the two creators of the ENIAC* (considered the first modern digital computer): J. Presper Eckert** and John W. Mauchly.** Maurice V. Wilkes,** one of the most important builders of British computers in the 1940s and early 1950s, received the award in 1968. Grace M. Hopper,** who helped to advance the use of higher level languages in the United States from the 1940s through the 1960s and who had perhaps been the first female programmer when she worked on Harvard's Mark I during World War II, won the award in 1970. Seymour R. Cray (father of the supercomputer that carries his name) was recognized in 1972.

For further information, see: American Federation of Information Processing Societies, *AFIPS and Its Constituent Societies* (Reston, Va.: AFIPS, Inc., 1985) and its *AFIPS Press, 1985 Publications Catalog* (Reston, Va.: AFIPS, 1985); I. L. Auerbach, "American Federation of Information Processing Societies," in Anthony Ralston and Chester L. Meek, eds., *Encyclopedia of Computer Science* (New York: Petrocelli/Charter, 1976): 56–57; *Annals of the History of Computing* 8, no. 3 (July 1986).

AMERICAN SOCIETY FOR INFORMATION SCIENCE (ASIS). ASIS was founded on March 13, 1937, as the American Documentation Institute (ADI). It was one of the first professional societies dedicated to providing services for those handling information, such as librarians. Many of its earliest members were organizations and government agencies primarily in scientific fields. Its early focus was on the use of microfilm as a learning aid. ADI developed microfilm readers and cameras and sponsored research on the general subject of microfilm. It also established programs for microfilming journals of interest to its membership.

In 1952 ADI expanded its membership base, becoming a truly national society for those interested in the general subject of information science in the United States. Although the original concept was heavily oriented toward library science, its membership developed a strong interest in the general subject of information science during the 1950s and 1960s. Because of this growing interest, ADI changed its name to the American Society for Information Science, a change that became effective on January 1, 1968. As one ASIS brochure stated, it was now "concerned with all conceptual, technical and practical aspects of the information-transfer process." This nonprofit organization subsequently published on the subject, ran seminars, and drew membership from all fields of information gathering, ranging from scientific institutions, publishers, and government agencies to data processing companies. By the 1970s the mission of ASIS was "to foster and lead the advancement of information science and technology," which was simply a modernization of ADI's original purpose.

By late 1985 ASIS claimed a membership of nearly 4,000 individuals and 150 institutions, making it one of the larger such organizations within the data processing community in the United States. There were thirty student chapters at various colleges and universities. A large number of its members worked with online information retrieval and library information networks. Others maintained data banks or worked with telecommunications, copyrights, machine translation of languages, or computational linguistics, all primarily within academic settings.

ASIS's publishing program has varied over the years; however, by the mid-1980s it had two bimonthly publications: *Journal of the American Society for Information Science* (JASIS), which carries serious research articles; and the *Bulletin*, which offers relevant news. ASIS has also published books and short monographs on topics of interest to its membership. The *Annual Review of Information Science and Technology* (twenty volumes published as of 1986) focuses primarily on advances in applications and research. Its *Proceedings* documents annual meetings (twenty-two volumes as of 1986).

For further information, see: ASIS, *Bulletin* and *Proceedings*.

AMERICAN TELEPHONE AND TELEGRAPH (AT&T). Best known as the largest supplier of telephone services to the United States in the twentieth century, AT&T has always been a major participant in the development of transmission technologies, communications, computers, and related devices, and as the owner of Bell Laboratories,† the source of important computer developments of the 1930s and 1940s and still a major research center. Beginning in the 1920s, AT&T conducted considerable research on switching systems (later called relay computers) to move telephone conversations around the nation and later to transmit or transfer from various points data by way of its Bell System. In 1983 it was already the twentieth largest data processing vendor as measured by data processing revenues. Many data processing industry watchers expect this multibillion dollar company to become an even greater force within the industry before the end of the 1980s.

In 1877 Alexander Graham Bell (1847–1922) patented the telephone. In order to market his creation, Bell agreed to accept financial assistance from a group of businessmen from Boston who, within two years of their arrangement with Bell, had taken control of his telephone patents. They had formed a company called the Bell Telephone Company and then renamed it the National Bell Telephone Company; finally, it became known as the American Telephone and Telegraph, or simply AT&T. The original board of directors, in order to raise funds to manufacture and sell the telephone, sold the J. P. Morgan and Company a bond for $150 million which allowed Morgan to gain control over the phone company. Throughout the 1880s and 1890s, the firm squeezed out competition by acquiring patents or through legal battles over patent rights. By 1906 AT&T was also concerned about its lack of in-house technical expertise to develop new products, particularly for the possible development of radio communications. Therefore, investments in research and development began.

By 1915 AT&T was offering long-distance telephone service. During World War I, the company conducted military-related research for the U.S. government in radio (wireless), telephonic communications, and radar, projects that were continued into the 1920s. Battles over who should control radio patents and communications resulted in an agreement in 1920 that in effect divided communications into two groups: radio and telephone. Under this arrangement, radio communications would be provided in the United States by General Electric,† Radio Corporation of America (RCA),† Westinghouse, and some other companies. They would share necessary patent rights, while the U.S. government would be confident that technological trade secrets would not be exported overseas. Telephone communications would be offered only by AT&T and Western Electric. This agreement heralded AT&T's march into a new world of regulated telephone services, which in turn nurtured the development of the nationwide Bell System. By the early 1960s, Bell would control over 80 percent of all telephone service in the United States. Disputes continued over policies and patents throughout the 1920s, however, as a result of which the government established the Federal Communications Commission (FCC) in 1934. The birth of the FCC formalized the establishment of AT&T as a regulated monopoly.

Earlier, during the 1920s, technical developments within the AT&T community were organized into the Bell Laboratories. Established at Princeton, New Jersey, in 1924, its mission was to develop necessary technologies to support nationwide telephone services. As part of that exercise, scientists at Bell Labs developed equipment in the 1920s and 1930s that could automatically switch telephone calls from one phone to another using automated switching equipment. This technology involved translating voice communications into electrical impulses, carrying these voice communications through telephone lines over long distances, then switching them by relay devices, and finally retranslating them into voices at the other end of its lines. The devices created to handle such communications during the switching process were computers of sorts (often called relay computers by World War II). These devices had to deal with issues concerning later computers: input of data that had to be converted to digital form and into electronics, manipulation of such data quickly and correctly, storage of information, and input/output. Thus, one of the roots of the modern computer can be identified as the work done at the Bell Laboratories during the 1920s and 1930s.

AT&T's Bell Labs became one of the most important research centers in the United States by the start of World War II. By the end of the war, it was recognized as a national treasure, a reputation it still had forty years later. During the early 1940s research and development on computers for government projects (mainly military) and for the improvement of telephonic communications proceeded at a rapid pace. By the end of the decade, Bell Labs had developed nearly a half dozen different models of its relay computer and had installed several in various government agencies.

The 1940s were capped with a major technological breakthrough for the computer industry in particular and the electronics industry as a whole, when scientists at Bell Labs developed the transistor. It was first demonstrated within the laboratories on December 23, 1947. This crude, germanium-wire unit, however, had no moving parts, could amplify sound, and used no vacuum tubes that burned out too frequently. It did, with more refinement, replace the vacuum tube by the late 1950s and early 1960s. The impact was profound because a transistor was a hundred times smaller than a vacuum tube, significantly more reliable, and less expensive. It was one of the most important technological breakthroughs of the twentieth century. It made possible the growth of electronics worldwide, paving the way, for example, for lightweight, small electronic devices such as radios and alarm clocks and, later, rockets. Similar results were later evident in the expansion and results of research and development of more reliable, less expensive computers with greater capacities and capabilities during the 1950s and 1960s. The world was told of the transistor on June 30, 1948. During the 1950s, this technology was gradually imbedded in various switching devices made by AT&T as it overcame the problem of how to manufacture transistors cost-effectively and to make them even more efficient. Quite unlike the past, when it patented and hoarded its technical developments, the company shared its transistor technology with the scientific community.

Several factors account for this change of strategy. First, in 1949 the U.S. government filed an antitrust suit against AT&T in which the Justice Department sought to divest AT&T of its manufacturing facilities of Western Electric. Lawsuits to protect patent rights on the invention of the transistor might enhance the government's case that AT&T was acting in a monopolistic manner and was too big within the communications industry. Negotiations with the Justice Department suggested the significance of AT&T's role at that time within the fledgling computer industry. In the consent decree of 1956, the company agreed not to sell either transistors or computers in the open market in exchange for retaining possession of Western Electric. It could also continue to develop computer-related equipment for its own internal use and to satisfy military contracts. Meanwhile, Bell could and did sell licenses to other companies for related patents to the transistors. The fact that such an important part of the consent decree of 1956 related to AT&T's role in computer technology was an early indication of its increasing importance in the new industry.

Because AT&T had considerable experience with computer-related technologies at the start of the era of computers (1940s–1950s), it had the potential of becoming a major force in the new industry. In 1950, for example, it was the largest company in the United States, valued at more than $11 billion. It had the massive manufacturing facilities of Western Electric which were built to produce electronic equipment and had been in existence for over two generations. AT&T's power was enhanced by its possession of Bell Labs, already one of the most important research facilities in the world. If one were to look at Western Electric by itself, which manufactured telephone equipment and computer devices for AT&T's use and to fulfill military contracts, it was large. In 1950 it sold $758 million worth of products, making it one of the largest manufacturing companies in the United States. It was larger than International Business Machines Corporation (IBM)† or any other company then capable of producing computers. AT&T's critical weakness in 1950 and for the next generation was its lack of an outstanding marketing organization capable of selling computers in a competitive environment. This weakness would make or break the company in the second half of the 1980s.

Thus, on balance, AT&T's strength was in the area of technological development at the dawn of the computer era. Even as early as the 1930s and early 1940s, Bell Labs was consulted on specific computer-related projects. Thus, for instance, it provided consultation to J. Presper Eckert** and John W. Mauchly** when they were building the ENIAC* during the 1940s, and later to other major projects throughout the late 1940s. The invention of the transistor brought Nobel Prizes to three members of Bell Laboratories. In the early 1950s the company had its own time-sharing computer system in operation, and, in 1954, it announced TRADIC, a general-purpose digital computer.* The fact that AT&T did not become the giant of the computer industry was due largely, of course, to the consent decree signed in 1956.

Yet under the terms of that agreement, it could continue to sell such products to the government. In conjunction with permission to develop equipment for use within the Bell System, AT&T has continued to do research and development until the present, primarily through its facilities at Bell Labs, making the company a continuous source of innovations. Moreover, it meant that in terms of technological capability within the company, AT&T could reenter the data processing industry in the 1980s with the same awesome capabilities it had in the early 1950s. This expertise extended to all kinds of data processing technologies: computers, peripheral equipment, terminals, modems, and software (both operating systems and application programs).

The restrictions imposed on its data processing operations in 1956 did not mean that sales from data processing products would necessarily remain small. In fact, the opposite was the case. In 1952 data processing revenues within the U.S. market amounted to $770,000, and by the end of 1963, these had grown to over $97 million, making it the third largest computer vendor in the industry after IBM and Sperry Rand.† In 1963 sales of computers to U.S. government agencies alone accounted for approximately $32 million. Throughout the 1960s, data processing sales continued to grow. Between 1964 and the end of 1969, revenues from sales of data processing products within the United States rose from $125.6 million to $477.75 million. Sales from Western Electric were to the Bell operating companies, particularly for telephone switching equipment. Sales to U.S. government agencies were another source of revenue. For each year during the 1960s, these averaged just over $50 million. AT&T competed against IBM during the 1960s, primarily for sale of terminals and message switching processors and for government customers or operating companies within the Bell System such as Southwestern Bell.

Direct dialing capabilities in the 1960s drove up the demand for additional, more sophisticated switching equipment, such as the No. 1 ESS system. Sales to the operating companies skyrocketed during the decade—from $20.4 million in 1964 to $62.4 million in 1966, to $108.5 million in 1968, and to $227.3 million in 1969. Modems also provided significant revenues. Modems are the devices that allow terminals to be attached to a telephone network to transmit data from one location to another. Sales of modems crept up during the decade from a low of $5.9 million in 1964 to $48.8 million in 1969—reflecting the significant growth in demand for distributed processing (both real-time and remote job entry) taking place throughout the industry. These modems were also some of the best available. Western Electric owned a subsidiary called the Teletype Corporation which in these years sold terminals. Its sales were also brisk: in 1964 it sold $62.4 million in products; its 1969 revenues were $110.7 million.

Additional products that appeared during the 1970s increasingly pushed AT&T deeper into the mainstream of data processing. This came at a time when data processing and communications were merging closer, creating a situation whereby technology and circumstances in the industry propelled AT&T and other communication vendors into data processing. This pattern first manifested itself with products. In 1975 the company announced its Dataspeed 50 terminals which went into direct competition with the industry standard: the IBM 3270 CRT. The Bell operating companies bought considerable numbers of AT&T's terminals, sometimes thousands. Also in 1975 AT&T announced the Dimension Private Branch Exchange (PBX) system. This widely used product considerably improved switchboards in hotels, offices, and universities while adding additional services, such as inventory control and wake-up calls. More popular functions included call-forwarding and intercom capabilities. The system was a computer, all solid-state and very well accepted as an industry standard. The Dimension PBX system was continuously enhanced throughout the late 1970s and early 1980s. Transaction Network Service (TNS) appeared the following year to manage large volumes of computerized commercial transactions using terminals and message switching.

Perhaps the most long-awaited AT&T product of the 1970s was the Advanced Communications Service (ACS). It had been under development for a long time, but the company finally submitted a tariff request to the FCC to offer ACS as a product. ACS would offer computers and other hardware from AT&T to manage and control networks of computers, terminals, and disk storage devices belonging to its customers. ACS was remarkable in its capacity both to perform network communications functions and to execute normal data processing applications. It was also seen as a direct competitor to IBM's System Network Architecture (SNA), which was announced before ACS and had considerable software available by the early 1980s. ACS put AT&T in the middle of the distributed processing area of the data processing industry.

For that industry, the 1970s were marked by a number of issues affecting all participants, not the least of which were technological developments and the U.S. government's antitrust suit against IBM. Of nearly equal significance, however, was AT&T's relations with the federal government, particularly at the end of the 1970s and early 1980s. AT&T's role within the data processing industry was most profoundly affected by decisions made both by the FCC and the government concerning the data processing community in general.

The chronology of AT&T's relations with the government is a short one. On November 20, 1974, the government filed an antitrust suit against AT&T and its subsidiaries and operating companies, and proposed the breakup of the Bell System as a remedy. On September 11, 1978, the judge in the case, Harold Greene, introduced a schedule for pretrial depositions. The trial itself began on January 15, 1981; by this time the data processing industry was in the midst of a major debate on the role of AT&T within the industry. On January 8, 1982, the government and AT&T came to an agreement, settling the case which was accepted by the court on August 11, 1982. During that fall, AT&T submitted plans to the court describing how seven area operating companies would be created independent of AT&T, and Judge Greene accepted the plans on July 8, 1983. The breakup was formalized effective January 1, 1984. These events effected a profound change within American industry generally and the data processing community specifically.

The technological realities of the 1960s had produced certain pressures during the 1970s. The cost of computers was dropping rapidly, but the expense of telephone service was declining more slowly. One still had to go through the Bell System at a time when an increasing amount of data was being communicated. Pressure mounted from within the data processing industry for other firms to provide more reliable and less expensive technologies to transmit data. This demand directly threatened the phone company's monopoly. Increasingly, the battle lines were being drawn, with IBM wishing to move into the communications field and AT&T wanting to do more with data processing. The FCC also saw the need to rethink its regulatory mission and specific policies regarding communications. AT&T's role was not insignificant. It had a payroll of 1 million people, owned twenty-three operating companies, and in addition had Western Electric, Bell Labs, and some other smaller companies. IBM was growing in size and by the late 1970s was in the top ten of the *Fortune* 500 list. By the early 1980s, it was projecting revenues of over $100 billion by the end of the decade. Data processing and telecommunications, two distinct industries, were blurring into one new industry concerned with worldwide networks, technical options for implementing such communications systems, electronic mail, automated offices and plants, and processors buried into every form of machinery and function imaginable. Even in the home the impact was evident. Gradually, individuals wanted to buy their phones, were using microcomputers tied into networks, and were shopping around for less expensive long-distance service.

The government's role was a lengthy one marked by major considerations. The Interstate Commerce Commission had regulated long-distance phone service as early as 1910. In 1934 the FCC came into existence to impose order on U.S. communications policies. In the early 1960s the FCC was forced to give additional thought to telecommunications when the U.S. Congress considered the

Communications Satellite Act. That discussion opened a pandora's box of complaints about the Bell System, particularly from the manufacturers of computers and important telecommunications users within the data processing industry. Was AT&T holding back technological improvements? Could it provide better service or was it too big? Was competition possible and useful? Was the introduction of new technologies in the 1960s making it more reasonable to offer the public choices for telephone services? Were the reasons for having a monopoly in the first place no longer relevant? These and many other questions began to stir a debate within the industry and at the FCC. At the same time, AT&T was not allowed to participate in the traditional data processing industry because of the 1956 consent decree. Did that continue to make sense in light of recent technological innovations?

The FCC fully involved itself with these issues in an attempt to redefine policies. Finally, in 1972 it issued a new decision. The FCC argued that the data processing industry should not be regulated by the government. It further ruled that, although AT&T could not offer data processing services, other common carriers (such as Western Union) could. Meanwhile, pressure had been mounting to allow other firms to tie into the Bell System with their own equipment, such as with devices that allowed radio telephones to communicate through the Bell network. AT&T resisted these attempts and lost. Competition for Bell's services was allowed. Thus, MCI, the most widely known carrier competing for long-distance service, could rival AT&T, drawing even more attention in the late 1970s to the need for yet more revisions of old rules. During the 1970s the FCC increasingly began to believe that AT&T could no longer provide the most efficient services or introduce new technologies as effectively as in the past. Hence, arguments in favor of monopoly were diminishing in the face of growing demands to implement more cost-effective and superior technologies. At the same time, the Department of Justice had filed an antitrust suit against AT&T.

Then, in May 1980 the FCC issued an important statement, often referred to as its decision in Computer Inquiry II. Henceforth, it would be the policy of the FCC to regulate only essential transmission services that moved information and voice. Any additional services would no longer be regulated. These could include, for example, protocol management on a telecommunications line, application software transmission, or even the use and source of equipment in a customer's building. At the same time, the FCC said that AT&T could compete within all areas of the data processing industry without regulation, thereby in effect nulifying the consent decree of 1956. The FCC ruling, in short, said that other companies like IBM, Xerox,† and Exxon could compete in the communications field and that the large phone company could sell computers, terminals, peripherals, and software much like the traditional computer manufacturers. The telecommunications industry could now have many new entrants providing services during the 1980s. The more traditional data processing industry had a new competitor—AT&T—which could operate anywhere it chose in the industry.

AT&T now needed to change its marketing, moving from an era of monopoly and thus little competitive pressure to one fraught with high-tech competition. The company found itself in quite new circumstances. It therefore hired an ex-IBM'er who had had a successful marketing career within the giant computer company to head up marketing, Archie J. McGill. This bright executive came into the company with a mandate to reorganize marketing. He hired a number of ex-IBM'ers to help, and he set out in the late 1970s to redefine the company's marketing strategy. His efforts resulted in changes. First, came Advanced Communication Service (ACS) in 1978, which allowed a wide variety of customers to plug into a Bell network with a variety of equipment—AT&T's or someone else's. Earlier, only Bell's equipment had been allowed to be attached to AT&T's Bell System. Second, in that same year AT&T began a major education program for its sales personnel to teach them how to market within the new environment. Third, a new organization was being formed, modeled on IBM's marketing divisions. Called "Baby Bell," it had sales plans similar to IBM's marketing regions and branch offices, and many ex-IBM marketing representatives, marketing managers, and branch managers in the new organization.

The year 1978 was important for AT&T for yet another reason. That was the year that its chief executive officer (CEO), John De Butts, retired and was replaced by Charles Brown. The new CEO's mission was to change AT&T's regulatory mentality into one of a lean, aggressive marketing organization. He once said that doing that was analogous to a captain trying to change course with an oil tanker; it was a slow process, he emphasized. Brown created Baby Bell, staffed it with IBM'ers, and gave it assets worth $12 billion. Brown also pushed hard for a settlement of disputes with the government while preparing the company for battle against IBM in the communications and data processing arenas of the 1980s.

On January 1, 1983, American Bell (Baby Bell) formally came into existence, headed by Archie McGill. Its formal mandate was to sell existing AT&T products and to help define new requirements and work with Bell Labs to develop products and get Western Electric to build them. Although the number of products available for sale at the time was small, it did have PBX and, soon after, additional terminals. Within six months, however, McGill had lost an internal battle for control of Baby Bell's policies and was therefore gone. Then on January 1, 1984, "Ma Bell" was broken up, ceasing to be the AT&T empire that it had been for decades. Seven independent regional companies came into existence which could elect whether or not to continue doing business with AT&T. Judge Greene also allowed the seven companies to use the bell symbol which had belonged to AT&T and ordered that the parent company no longer use it. AT&T adopted a worldlike sphere as its symbol. The Bell System had changed. AT&T continued to own Western Electric and Bell Labs and to offer long-distance services but without a monopoly over it.

In the early months of 1984 the degree to which AT&T's phone monopoly had been broken up was becoming evident. Northern Telecom, Inc.,† a competitor of AT&T's Western Electric, built and sold to the new telephone companies digital switching equipment which allowed hundreds of thousands of phone calls to be carried from city to city. Western Electric's products were not competitive at the time, causing Northern Telecom to experience a growth in business in 1983 and 1984 of over 20 percent compound. In mid–1984 a regional telephone company announced that it would buy long-distance services from MCI while most of the regional companies were already acquiring telephones from a variety of producers and not just from Western Electric. The phone monopoly had indeed been broken up.

The changes caused AT&T's sales to dip. In 1983 revenues had been $69.4 billion, representing a growth over 1982 of 7 percent. Net income declined from $1.5 billion in 1982 to a net loss of $4.9 billion in 1983. Some of that loss was due to the company writing off assets at a faster rate of depreciation than it normally had done in the past as a means of reducing its long-term overhead expenses. Data processing revenues for the Unix operating system and other software were flat in 1983, generating approximately $75 million. PBX sales in 1983, however, did grow 4 percent over 1982, not an impressive record in light of the fact that communications sales for the entire industry expanded by over twice that percentage in 1983. Sales of terminals fell, as did income from maintenance services. Much of AT&T's lost business was picked up by regional companies. Thus, in 1982, 62 percent of all sales went into the Bell System's twenty-three companies. In 1983 this percentage rose to 71 percent. Hence, sales into what had once been the AT&T empire as a whole were still good, despite the significant bite taken out by competition. Despite all the changes of the early 1980s, AT&T ended 1983 as the twentieth largest data processing company in the United States. Industry analysts expected AT&T to continue developing data processing products to compete against the industry's major suppliers. An indication of this activity came in the spring of 1984, when AT&T announced that it would introduce a microcomputer to compete with IBM's Personal Computer (PC).

For further information, see: Jeremy Bernstein, *Three Degrees Above Zero. Bell Labs in the Information Age* (New York: Charles Scribner's Sons, 1984); Ernest Braun and Stuart MacDonald, *Revolution in Miniature: The History and Impact of Semiconductor Electronics* (Cambridge: Cambridge University Press, 1978); N. R. Danielian, *AT&T: The Story of Industrial Conquest* (New York: Vanguard Press, 1939); Franklin M. Fisher et al., *IBM and the U.S. Data Processing Industry: An Economic History* (New York: Praeger Publishers, 1983); Katharine D. Fishman, *The Computer Establishment* (New York: Harper & Row, 1981); Dirk Hanson, *The New Alchemists: Silicon Valley and the Microelectronics Revolution* (Boston: Little, Brown & Co., 1982); B. D. Holbrook and W. S. Brown, *A History of Computing Research at Bell Laboratories (1937–1975)*, Computing Science Technical Report No. 99 (Murray Hill, N.J.: Bell Telephone Laboratories, 1982).

APPLE COMPUTER, INC. Apple became a dominant producer of microcomputers from the day of its founding in 1977. Along with International Business Machines Corporation (IBM),† it shared the bulk of the small computer market and shaped its characteristics by the early 1980s. From 1977 to the present, Apple Computer has also been one of the most widely recognized computer companies in the data processing industry. The introduction of its Apple computer launched the microcomputer revolution which, by the mid–1980s, resulted in the sale of several million such devices in the United States alone delivered by over 150 different companies. The introduction of microcomputers that were easy to use and inexpensive expanded the use of computing as dramatically as had the introduction of the S/360* by IBM in the mid–1960s.

The two most important founders of the company were Steve Jobs** and Steve Wozniak,** both of whom had tinkered with electronics while growing up in California in the 1960s and 1970s. By the mid–1970s they had teamed up to build microcomputers in the garage of Jobs's parents in Los Altos, using spare parts primarily from Hewlett-Packard.† In the early 1970s one obtained a personal desktop computer by building it oneself; very few were commercially available. Jobs and Wozniak built such machines for members of the Homebrew Computer Club† (which met at Stanford University) and subsequently constructed fifty for the Byte Shop in Mountain View. At an early date Jobs wanted to develop a company, whereas "Woz" (as he was best known) became the technical brains behind what became known as the Apple I. They were joined by a veteran of the industry, A. C. "Mike" Markkula, then a thirty-eight-year-old "retired" millionaire and an ex-employee of both Fairchild Semiconductor Corporation† and later Intel,† to help form the company. At the time Jobs was only twenty-one years old.

While the company was being established, Wozniak designed the Apple II, completing it in 1976. This machine encouraged investors to provide the funding necessary to create the Apple Computer Company. More importantly, this microcomputer opened a new chapter in the history of data processing. This twelve pound machine was easier to use than any previous micro. Ultimately, after several years of considerable success with the Apple II, IBM entered the market in 1981 with its own Personal Computer (PC). By the mid–1980s, a multibillion dollar market had developed just for microcomputers.

The Apple Computer Company was formally established on January 3, 1977, and by the following January had a value on paper of $3 million. When Markkula joined with Jobs and Wozniak to form Apple, he obtained $250,000 in credit from the Bank of America, bought a third of the firm himself for $91,000, and collected an additional $660,000 from other investors to launch the company.

Its first product was the Apple II which Stan Augarten has dubbed the "Volkswagen of personal computers." Quite possibly, Wozniak himself initiated this analogy with the Volkswagen Beetle which characterized what he thought would be the final assessment of the Apple II. The machine had 16K of memory and was list priced at $1,195. It was introduced at a time when the demand for a simple desktop computer had been established by other vendors such as Osborne,† and when there was growing awareness and requirements for computing power outside of traditional data centers. The net result was a highly successful product which caused the Apple Computer Company to grow very rapidly. In 1977 sales were $775,000, and by the end of 1984 they approached $1.5 billion. It entered the roster of the *Fortune* 500 within five years of birth—the fastest any firm had ever done that. Growth was particularly rapid in its early years. From its sales of $775,000 in 1977, it closed the books four years later at $335 million. In December 1980 the company went public with shares selling at $22 each (8 percent of the net worth of the firm), giving the company a value of $1.3 billion. At that time Wozniak's shares were worth $88 million, Jobs's $165 million, and Markkula's $154 million. The company had grown from three employees to 4,700 by January 1984.

Although Jobs, Wozniak, and Markkula were the driving forces behind the launching of Apple Computers, other executives broadened the firm as it grew. The first president of the company was an experienced manager from National Semiconductor, Michael Scott. In 1981 Markkula took over the administrative reins as chief executive officer, and in 1983 the company hired the president of Pepsi-Cola, John Sculley, to fill the job. Jobs remained chairman of the board and active in each of the firm's major decisions. Wozniak headed up the company's development of new products. Under these and other managers, sales and profits rose. Sales topped $1 billion by the start of 1983, and in that year profits reached $77 million—a star performance within the data processing industry. These sales were supported by the Apple II, the workhorse of the company, but was also helped by Lisa and, in 1984, the Macintosh. By September 1980 over 130,000 Apple IIs had been sold. Similar statistics would prove the attractiveness of the Apple II in the following four years.

In 1981 IBM introduced the PC, providing Apple with considerable competition. By the mid–1980s the PC dominated the market for microcomputers. Most competitors and manufacturers of microperipherals and software built products compatible with the PC, giving IBM an important marketing advantage over the Apple. In late 1981, however, many industry pundits noted that, despite Apple's past successes, it was IBM's entry into the market that gave the microcomputer "legitimacy" and hence contributed to the dramatic rise in the use of such devices in the following years. Such logic did not diminish Apple's impact.

Apple's technology was impressive for its day. Lisa, for example, had introduced considerable technological innovations that were not then available on the Apple I or II, such as extensive graphics and a "mouse," that is, a hand-held device causing commands to be executed when pointed at various functions on the TV-like screen. Lisa, introduced in 1983, did not sell well. Competition from a series of IBM products along with a too-high price tag on Lisa of nearly $10,000 did not help Apple. IBM sold its micros primarily to its business clients whom Apple had failed to influence effectively. In 1983 this stiff competition, along with that of other vendors, affected Apple's financial performance. Apple's final quarter that year showed income falling to $5.1 million against sales of

$273.2 million as compared to earnings of $18.7 million on sales of $175.2 million during the last quarter of 1982. Apple's share of the microcomputer market decreased from 29 percent to 23 percent, while IBM's went from 3 percent to between 22 and 28 percent (depending on which source was cited). Meanwhile, Osborne, one of the first important microcomputer companies, declared bankruptcy.

The Apple Computer Company brought out additional, lower cost models of Lisa, and continued to sell successfully the Apple II and the Apple IIe (which had more memory and a better keyboard, and used less chips* than the II). The Apple IIe, introduced in January 1983, did well, it was purchased by 300,000 customers by the end of that year. For a brief period it had been named the Apple III, but later, when sales momentarily dipped on the new machine, the firm relied on the II as a more recognizable name.

The company needed a new product and thus focused its attention on what became known as Macintosh. It was introduced on January 24, 1984, following six months of hype and industry anticipation of a revolutionary micro along with a hostile advertising campaign aimed at negating IBM's products. This portable computer (it weighed less than 10 pounds and came with its own carrying bag), while technically sophisticated, could not run most of the currently available business software packages and had a small memory. It was introduced without a letter-quality printer, and its word processor software could not handle long documents, making the machine less attractive for the large business market than had been suggested. Yet it received a good press. But the Apple II and IIe continued to support the firm as it had in 1983 when the II culled 97 percent of all Apple's sales. Meanwhile, in 1983 the company sold only 20,000 Lisas and was destined to experience less than anticipated sales of the "Mac" in 1984. Even so the Macintosh was the product through which the company hoped to regain its earlier luster. It handled graphics, had the mouse which could be used instead of a keyboard, but was also priced at $2,500—too expensive for home use and too underpowered for commercial customers.

Although its last three products were not as successful as its first two, Apple Computer has remained a sizable force in the microcomputer market in the United States as of 1986. Its main contribution was to launch the microcomputer revolution—and it was nothing less than a revolution in computing—while it served as the classic example of the Silicon Valley company that did well with spectacular growth in a short period of time. Despite its brilliant success, however, like other one-product companies it could face a dark future. The history of Apple is also that of a company that had to develop complex organizations to support rapid growth while attempting to thrive in a highly competitive industry where technological leap-frogging was a constant threat. It was an environment in which even established firms, such as IBM, could show the same flexibility and business prowess when entering a new data processing market.

For further information, see: Stan Augarten, *Bit by Bit: An Illustrated History of Computers* (New York: Ticknor & Fields, 1984); Paul Freisberger and Michael Swaine, *Fire in the Valley: The Making of the Personal Computer* (Berkeley, Calif.: Osborne/McGraw-Hill, 1984); Doug Garr, *Woz: The Prodigal Son of Silicon Valley* (New York: Avon, 1984); Robert Levering et al., *The Computer Entrepreneurs: Who's Making It Big and How in America's Upstart Industry* (New York: New American Library, 1984); Stephen T. McClellan, *The Coming Computer Industry Shakeout: Winners, Losers, and Survivors* (New York: John Wiley & Sons, 1984); Michael Moritz, *The Little Kingdom: The Private Story of Apple Computer* (New York: William Morrow, 1984).

ASIS. See AMERICAN SOCIETY FOR INFORMATION SCIENCE

ASM. See ASSOCIATION FOR SYSTEMS MANAGEMENT

ASSOCIATION FOR COMPUTING MACHINERY (ACM). This is the largest and most important organization within the data processing industry. Historically, it has appealed to all segments of the industry from computer scientists (its original group of supporters) to end users, data processing managers to operators, programmers, and especially systems analysts. It has published the most important journals in the field and has served as a single voice for the industry. The ACM is also one of the largest industry-related associations in the United States. Traditionally, its president is an important figure within the industry.

ACM was established on September 15, 1947, at Columbia University as the Eastern Association for Computing Machinery. Its purpose was to bring together scientists and engineers interested in computers. In September 1949 it adopted a constitution, and in December 1954 it was incorporated in the State of Delaware as the ACM. Its first several presidents were important developers of computers in the United States. John H. Curtiss** served as the first president, and John W. Mauchly,** co-creator with J. Presper Eckert** of the ENIAC* computer, held the position from 1948 to 1952. Harry D. Huskey,** another important computer scientist, was president between 1960 and 1962. Other notable scientists within the industry who were president included Allan J. Perlis**(1962–1964), Anthony Ralston (1972–1974) who co-authored a massive dictionary on data processing, and Jean E. Sammet**(1974–1976), developer of COBOL* and author of an important history of programming languages.*

Local chapters have been established in the United States and around the world, each reporting to regions and then to the elected governing body of the ACM. Technical programs cover every single aspect of data processing from hardware to software developments, their use, and the role and education of professionals within the industry. In 1961 the ACM launched its National Lectureship Series to make important speakers available to local chapters. In 1966 it established the Turing Award, which is given annually to a person who makes an outstanding technical contribution. This award has traditionally gone to recognized life-time contributions, such as building new computers or developing advanced programming languages. Some of the recipients have included Perlis, Maurice V. Wilkes,** Marvin Minsky,** Allen Newell,** and Herbert A. Simon.** In 1970 the Distinguished Service Award was created, and throughout the 1970s other types of recognition were established. ACM's experience demonstrates that the data processing industry had established a self-identity by the 1960s, had organized itself around various issues, and had emphasized values that were reflected in the actions of this and other associations.

Almost from its beginning, the ACM produced publications of interest to its members, especially those involved in research and teaching. The *Journal of the Association for Computing Machinery* was launched in 1954 for the purpose of providing technical papers on research and development. *Communications of the ACM* was started in 1958 as a monthly publication also dedicated to technical issues. It has always been one of the best sources of information on the development and use of specific programming languages and other software tools. Almost from its inception, *Communications* became the most important technical journal on all aspects of data processing's technologies. In 1960 ACM began publication of *Computing Reviews* as a monthly survey of publications concerning data processing. Subsequently, various bibliographies have also been published. *Transactions on Mathematical Software*, a quarterly, began in 1975, focusing on theoretical and applied issues concerning mathematical software and algorithms for computers. The ACM has also issued dozens of other collections of proceedings and transactions of meetings concerning a wide range of technical issues.

ACM holds annual conventions, although in recent years more than one have been conducted along with ongoing seminars and regional conferences. ACM has also associated itself as a founding member with the American Federation of Information Processing Societies (AFIPS)† and has participated in each National Computer Conference (NCC).† By early 1983 membership in ACM exceeded 60,000.

For further information, see: I. L. Auerbach, "Association for Computing Machinery (ACM)," in Anthony Ralston and Chester L. Meek, eds., *Encyclopedia of Computer Science* (New York: Petrocelli/Charter, 1976): 128–129; and a series of ACM publications: *ACM Publications: Copyright 1947, 1985 and 1995!* (New York: ACM, 1985) and *1985 Publications Catalog* (New York: ACM, 1985).

ASSOCIATION FOR EDUCATIONAL DATA SYSTEMS (AEDS). This organization provides education through the use of data processing technology, and it encourages the development of new techniques in the use of computers in education and disseminates information on the subject. Its establishment in 1962 represents early recognition of the role computers could play in education. Founded when computers were just beginning to be important outside of large government agencies, businesses, and universities, AEDS was one of many organizations created to promote the use of computing in new ways.

AEDS was founded by a group of professional educators and technically trained individuals, and incorporated in the State of Florida. It continuously carried out its mission with workshops and seminars on programming languages,* the use of video disc systems, the evaluation of computer courseware, word processing, and the management of computerized applications related to education. Each year it holds a convention in which papers are read on relevant subjects. Its major publications include the *AEDS Journal* which carries articles on research and experiences using computerized technology, *AEDS Monitor* for bimonthly dissemination of short articles on computers and education, and the bimonthly *AEDS Newsletter* which presents news on the organization and related issues to its members. AEDS also publishes books and pamphlets on such subjects as the use of computers in education, and on the role of microcomputers for instruction.

AEDS is headquartered in Washington, D.C., and is the oldest association dedicated to working with technology in schools. Its membership in 1985 consisted of 6,500 teachers, administrators, specialists in computer science, and individuals from industry and government. Most members are from the United States, but some also reside in Canada, Mexico, and Europe.

For further information, see: I. L. Auerbach, ''Association for Educational Data Systems (AEDS),'' in Anthony Ralston and Chester L. Meek, eds., *Encyclopedia of Computer Science* (New York: Petrocelli/Charter, 1976): 130.

ASSOCIATION FOR SYSTEMS MANAGEMENT (ASM). The ASM was one of many associations that emerged within the data processing industry to provide education for those working with computers. Its primary objective was to offer seminars and publications for those responsible for information resource management systems. It is one of the larger such organizations providing continuing educational programs within the industry.

ASM was founded in Philadelphia on May 23, 1944, by a small group of individuals working with computer systems. On June 1, 1945, it adopted its first constitution and set of bylaws and the name Systems and Procedures Managers Association. On September 12, 1946, this organization merged with another such group from New York to form the Systems and Procedures Association of America. This new organization was established at the dawn of modern data processing when government agencies, universities, and most notably companies were beginning to use computers in large numbers. This situation in turn created a new class of workers: those concerned with the establishment and maintenance of data centers for computer users. From this same group emerged the data processing departments which today are such an integral part of any organization and which are staffed with programmers, systems analysts, operators, and their management. It is this particular group of people, particularly the managers, to whom such associations as the ASM and the Data Processing Management Association (DPMA)* appealed.

As the data processing industry grew in size and self-awareness, so did the number of managers building careers as users of computing technologies. ASM established chapters around the United States and in Europe. In 1968 it adopted a new constitution and bylaws to support a reorganized and larger entity. It was then that the name was changed to the Association for Systems Management. Throughout the 1970s membership expanded and by early 1985 reached more than 9,000. Over 100 chapters exist in the United States, Canada, and in thirty-two other countries. Chapters report to twenty-one division councils, which in turn are under the control of a twenty-six member international board of directors.

The ASM holds an annual convention and monthly meetings at each chapter. It has also sponsored the *Journal of Systems Management* and over 1,400 publications in its Systems Reference Library series.

For further information, see: ASM: Association for Systems Management (Cleveland, Ohio: ASM, n.d.).

ATARI CORPORATION. Atari was the first company to produce computer-based video games; by the early 1980s it had become the largest manufacturer of such products in the world. The word "Atari" comes from the Japanese word meaning "you are about to be engulfed." The phrase was used in a game called Go which, like Chess, tested a player's skills in strategy. The statement "Atari" played the same role as "Checkmate" in Chess. Atari Corporation was founded in 1972 with some $500 in capital, and it exceeded $1 billion in sales in 1982. It thus represented one of the fastest growing companies in the world. Its products were initially coin-operated games in taverns, followed by versions that could be played at home. By the early 1980s its products included a small home computer. Atari, particularly in the United States, introduced video games based on data processing technology which were played by millions of people. It also killed the demand for pinball machines.

The company was established by Nolan K. Bushnell (1943–), an engineer from Utah. Following graduation from the University of Utah, Bushnell worked for Ampex in California (the developers of videotape) while at night he tinkered with microcomputers and games in an attempt to build one. He had first been exposed to arcade games while in college in a part-time job and to computers as part of his formal education. He had already seen games played using large mainframes, such as Space War, which had been in existence since the early 1960s when it was developed at MIT. But he wanted to develop a game that would fit into an arcade much like pinball machines. He succeeded, and by 1983 his company dominated the $3 billion market for arcade video games.

Bushnell built his products by taking advantage of declining costs in microprocessors and of improved electronics in the early 1970s. His first product was made up of a black-and-white television set, logic boards, transistors, diodes, and printed circuits. His idea was to display the playing field on the TV screen and to use the circuit boards to control the action. His first game, and the first commercial video game ever produced, was called Computer Space, which he completed in 1971. Spaceships fought flying saucers and used Newton's third law of motion closely: for every action there was a reaction. Through Nutting Associates he sold the game to taverns. Hardly 2,000 copies were sold, however, primarily because customers found it too complicated and the concept of an electronic video game was still too new.

During the following year (1972) Bushnell worked on a new game called Ping-Pong or simply Pong. Unlike the first game, it was very simple to play. The first copy was installed in Andy Capp's Tavern in Sunnyvale, California, and was an instant success. The story was frequently told that the tavern manager called Bushnell after the first day to complain that it did not work. Bushnell examined the machine and discovered that the milk carton used to hold the quarters charged for the game was overflowing, thereby choking the mechanism itself. A user fed the machine a quarter to enjoy temporary control over the computer playing Ping Pong. Carl Sagan later commented that it was a thinking person's pinball, arguing that with it "a player can gain a deep intuitive understanding of the simplest Newtonian physics." Most found the game simply fun. The machines were sold to distributors of pinball devices for about $2,500 to $3,000 each. The game killed the pinball industry because video equipment was easier to maintain and people played them more. By 1974 over 100,000 copies of the game had been sold. Yet only 10 percent came from Atari; the others were counterfeits from U.S. and Japanese sources. That same year Bushnell introduced another product called Gran Trak 10 for those who wanted a race-car driving game, but he lost a half million dollars on it.

Atari was officially formed in 1972 to market Pong. The first device was shipped in November 1972. During 1973 Atari sold 10,000 at $1,200 each. In 1975, a home version appeared under contract with Sears Roebuck & Company for 150,000 copies. Magnavox, which made television sets, had brought out a video game by Christmas 1972 that could be attached to a home computer. In effect, this opened up a new market—the home—for video games. Its first product was called Odyssey and sold 200,000 cartridges and peripherals in the first three years. Atari concluded that it, too, could market products for use in homes soon after Magnavox brought out its product. Atari's first introduction into the home market was a version of Pong. In order to obtain sufficient capital to fund the manufacture of this product, Bushnell approached various investors, but finally Sears agreed to the deal, gaining exclusive rights to the game during the fall of 1975. The retail giant offered it for sale in 900 stores. During the next three years Sears sold 13 million Pongs. In 1975 Atari enjoyed revenues of $39 million and a net income of $3.5 million—very handsome by data processing standards.

The following year Atari introduced Breakout. This game used a paddle to hit the ball against a brick wall; each time the ball hit it, a brick popped out until they all disappeared. The company sold 15,000 copies of this game. By the end of that year, the demand for video games in the United States alone appeared almost insatiable, and Atari did well as a leading supplier. Furthermore, its products were then being introduced into Europe as well.

In 1976 the only two important manufacturers of video games—Atari and Magnavox—could not satisfy the demand while competitors introduced their own products. Microchips became more sophisticated, capable of greater function, and their costs dropped. Fairchild Instrument and Camera Corporation (later Fairfield Semiconductor Corporation†) developed programmable chips* now used in such games. In 1976 the Radio Corporation of America (RCA)* introduced its own products, and during the following year, Bally, best known for its games used in gambling casinos, came out with its own video arcade games. Yet chip manufacturers who serviced this part of the data processing community could only satisfy approximately 50 to 60 percent of the demand for microprocessors in 1976.

As with software products and, later, personal computers, vendors began to introduce new games, leapfrogging each other's technologies and features, thereby heating up competition while placing pressure on companies such as Atari to bring out new products. The costs for devices and games also dropped. For example, the earliest releases of Magnavox's Odyssey 200 sold for $129 in 1975. The following Christmas these could be purchased for only $39. In order to raise more capital to fund additional research and manufacturing, Bushnell sold Atari to Warner Communications, Inc., in 1976. Warner believed that with the sale of films and records then declining it needed another lucrative product in the entertainment business. Video games seemed a logical source, particularly those in arcades stuffed with quarters.

Warner Communication's acquisition opened a new chapter in the history of Atari because the unstructured, casual appearance of Atari gave way to more formal controls and operations. For one thing, procedures were instituted for budgets, expenses, and reporting that had not existed when Atari was completely under Bushnell's control. For another, new management from Warner in Atari, such as Emanuel Gerard who negotiated the acquisition on behalf of Warner, wore business suits and ties and discouraged blue jeans and open collars. More seriously, Warner's management did not pamper or celebrate the engineers who invented games at Atari as had previous management. That change caused dissension, and in time many left either to work for other firms or to establish their own businesses.

But before and during the transformation from complete independence to control by Warner, other activities of note took place at Atari. By the early 1980s, Warner had invested $120 million in Atari. Meanwhile, the video game manufacturer had developed the Video Computer System (VCS) which consisted of game cartridges and color graphics that sold for $200 in the home market. This system was programmable and thus flexible. Cartridges were so designed that they cost about $10 to make and could be sold for up to $25; VCS promised to be a very profitable product. Atari brought it out in time for the Christmas shopping season of 1977, but at first experienced slow sales. There were production problems with the machine (such as static electricity knocking out controls for users), but other vendors had their concerns as well. Some of their games did not work correctly, and Atari was not immune from this problem either. At Atari, for example, Home Run, a baseball game, was written by someone who did not understand the game's rules. Thus, if a batter swung and missed a pitch he was not given a strike; rather, it was scored as a ball. That game was fixed. During 1978 the video game market was depressed for all vendors, including Atari. The firm's particular problem was that it continued to manufacture VCSs faster than they could be sold. The company was still dominated by engineers, not marketing personnel, and as a consequence Warner had to remedy the problem with managers who had backgrounds in marketing and sales.

Atari experienced considerable competitive pressure from the Japanese in 1979 when the latter introduced Space Invaders into the American market. The game sold well because it was the first readily available game controlled by a chip. That microprocessor coordinated the activities of alien tadpoles attacking a garrison that used movable laser cannon for defense. It was a complicated game to design and operate, challenging both its inventors and users, all of whom were ready for this additional level of sophistication. A total of 350,000 copies sold around the world, 55,000 in the United States alone. Atari responded with Starship and Super Breakout, both of which failed to unseat the Japanese offering. In 1980, however, Atari introduced Asteroids which brought back market share and profits.

As with many products introduced by members of the data processing industry, it was technological innovation that helped companies thrive. Prior to Asteroids, a player controlled games by moving a gun back and forth across a fixed axis. But now Atari gave users the ability to send their spaceships to any point on the screen. To do that, each game used two microprocessors. Atari sold 70,000 copies of the coin-operated game at a price of about $2,700 in the first year and in time ended up with 100,000 sold, making sales second only to Space Invaders. Atari took its profits and bought the rights to market the Japanese competitive product exclusively.

Sales in 1980 reached $415 million, with operating revenues growing fivefold over the previous year to $77 million. Atari's income was one-third of Warner's that year. In the early 1980s the move to increased animation in video games started, contributing to additional sales and revenues from newer products. For example, soon after it had become obvious that Asteroids would be a success, the company introduced Battlezone, a game involving tanks fighting each other. When a tank moved forward, the landscape advanced in the proper perspective. At the time users considered the game a major technological step forward; by the standards of 1985, when cartoon-quality animation could be found in video games, it appeared old-fashioned. But when it came out, it was leading edge, required greater complexity in programming, and used three microprocessors. Atari sold 25,000 copies at $3,000 each for use in video arcades. It was followed by other games that relied on the same state-of-the-art technology: Missile Command, Tempest, Dig Dug, and Space Duel among others.

The most important game to appear in 1982, however, did not come from Atari but rather from Japan: Pac Man. Midway Manufacturing Company, located in Chicago, was licensed to market it in the United States. The company made it the best selling video game within two years, relegating Asteroids to second place. For the first time, a video game character had personality (relatively speaking), whereas Atari's products were essentially games of marksmanship.

By the mid–1980s Bushnell was gone as a result of a dispute with Warner. Profits came from coin-operated arcade machines and VCS home products. Almost 75 percent of the home market in 1983 went to VCS. A third part of the company, established in 1980, sold personal computers but not very well. In 1980 that division lost nearly $10 million and in 1981 ran third behind Tandy Corporation†/Radio Shack and Apple† in units sold. As late as January 1986 it was advertising personal computers compatible with Apple's while the industry's standard had years earlier settled on International Business Machines Corporation (IBM).†

In many ways Atari's history reflects patterns common to the data processing industry. In the first place, like many other firms, it started with an entrepreneurially minded engineer's idea for a product. It did well, but that engineer found it difficult to successfully manage a marketing organization and hence was either swallowed up by a firm that did or simply went out of business. Humble beginnings seemed part of the legacy. In some cases personal computers were built in garages belonging to inventors' parents (Apple for example) or in children's bedrooms at night (Atari). The company's founder was young, typically under thirty years of age. In the case of Atari, Bushnell was an "old" man of thirty-four when he finally left his company as a multimillionaire. Success required new products that incorporated advanced technologies appearing rapidly (frequently every few months in some sectors of the industry) in what was a highly competitive market.

Atari spun off surrogates. If Fairchild was considered the parent of many high-technology companies in Silicon Valley, Atari played the same role for video games. For instance, in 1981 some key engineers left Atari to form their own software companies called Imagic and Activision. At the latter, creators of games were treated as celebrities rather than as faceless engineers. Thus, when a buyer acquired a game from Activision, included in the package was a picture of the designer. That company sold $15 million worth of games in its first year of operation and approached $50 million by the end of 1982. Imagic approached $10 million in the same year. Even Nolan Bushnell created a spinoff of his own company by forming Chuck E. Cheese Pizza Time Theater, a chain of restaurants that combines pizza, video games, and robots reminiscent of Disney's characters that entertain children. When he sold Atari to Warner, Bushnell made $15 million personally. In a press interview in mid–1982, he estimated his worth at $70 million. A whole new generation of video game manufacturers existed by the mid–1980s who had so decimated Atari either by resignations from that company or through their own products that in 1986 and beyond interest in Atari was largely historical.

For further information, see: Scott Cohen, *ZAP! The Rise and Fall of Atari* (New York: McGraw-Hill, 1984); Colin Covert, "Video Gamesmanship: The Rise and Fall and Rise of Atari," *Ambassador* 15, no. 8 (August 1982): 29–37; Everett M. Rogers and Judith K. Larsen, *Silicon Valley Fever: Growth of High-Technology Culture* (New York: Basic Books, 1984).

AT&T. See AMERICAN TELEPHONE AND TELEGRAPH

AUTOMATIC DATA PROCESSING (ADP). ADP is one of the most successful service bureaus in the history of modern data processing. A service bureau is a company that provides computing facilities or services for a fee. A common example of such a service would be the processing of a company's payroll, and the printing of checks for employees and reports for employers. ADP's history reflects developments within many service bureaus in the second half of the twentieth century. No history of the data processing industry would be complete without an appreciation of this aspect of its activities.

ADP came into existence in 1948 at the dawn of modern data processing in order to do payrolls, using traditional manual bookkeeping methods and accounting machines. It initially installed International Business Machines Corporation's (IBM's)† punched card equipment and in November 1961 its first computer, an IBM 1401.* In 1964 it joined thousands of other customers of IBM in ordering its first System/360.* By then it was the largest processor of payrolls in the United States, a position it maintained for the next two decades. In 1964 alone it had over 500 companies as customers and processed payrolls for over 80,000 people. From 1964 to 1969 it expanded its services and developed a successful marketing force that increased revenues twentyfold and earnings by fifty-eight times.

During the 1970s ADP's major applications included transactions for brokerage firms, time-sharing, and management of portfolios. Although services in its early years were concentrated in New Jersey, by the end of the 1970s it had facilities throughout the United States. In 1970 a total of 7,000 companies depended on ADP for processing payrolls amounting to $5 billion in wages. Growth in the company's revenues rivaled that of the most successful firms in the data processing industry in general. Revenues rose from $187,000 in 1957 to nearly $2 million in 1963. These jumped to $4.7 million the following year, to $20 million in 1968, and, in 1970, to $37 million. ADP expanded its services even further to include order-entry, billing, and inventory control. Terminal-based applications were added by which customers used Cathode Ray Terminals (CRTs) at their locations connected to ADP's computers via telephone lines to enter data and do inquiries. ADP developed a nationwide telecommunications network to support such services. Accounts receivable, sales and profit analysis, general ledger, and other accounting and financial applications were enhanced and were made interactive instead of just being processed in batch mode. By the late 1970s, ADP was also offering database applications.

Unlike most of its rivals in the service bureau industry, ADP also sold hardware built for it by other firms. In 1975 a minicomputer system which it marketed made possible distributed processing to ADP's own large computers. In the beginning (1975) the service was offered for applications in inventory control and accounting. In 1978 ADP brought out ADP/Onsite which was a service involving the use of a computer physically located in a customer's building to do processing while networking into ADP's data centers. It could be used to house several databases, four languages for programming, and applications from ADP. All of these new services made it possible for the company to continue its growth. In 1979 ADP had over 75,000 users cutting across all major industries: manufacturing, distribution, finance, insurance, and government. That year alone revenues reached $371 million. During the 1980s the company continued to thrive, expanding its offerings with decision-support systems and more database applications.

For further information, see: Franklin M. Fisher et al., *IBM and the U.S. Data Processing Industry: An Economic History* (New York: Praeger Publishers, 1983).

B

BELL LABORATORIES. Bell Labs has been one of the major centers of basic research in electronics and physics throughout the twentieth century. It has historically served as the primary research and development facility of the American Telephone and Telegraph Company (AT&T).† It has also been an important source of technology in the history of computers and data processing in general. At Bell Labs, for example, transistors were invented and important research was carried out on semiconductors, beginning in the 1930s. Bell Labs was home to seven Nobel Peace Prize winners, four of whom were in solid-state physics. Scientists there continued to produce remarkable discoveries in each decade and, most recently, during the 1980s, produced solid evidence that our universe was the result of a "big bang" explosion.

To appreciate the magnitude of what Bell Labs has meant to the history of data processing, a discussion of its recent past will be helpful. Prior to the breakup of AT&T in the early 1980s, Bell Labs employed nearly 25,000 people in thirteen locations in New Jersey and at another seven locales throughout the United States. After the divestiture (effective January 1, 1983), the labs transferred some employees to AT&T, reducing its population to 18,000. This national treasure boasted that at least one-third of its staff had Ph.D.s in the physical sciences. The main location, at Murray Hill, New Jersey, was constructed in 1941 and subsequently expanded. In the early 1980s it housed over 4,000 people conducting research in almost every field of science. Bell Labs is a subsidiary owned by AT&T; yet at one time 50 percent of it was owned by Western Electric, the manufacturing arm of the telephone empire. After January 1, 1983, Bell Labs remained the sole property of AT&T. By the early 1980s some 10 percent of the Labs' efforts were devoted to basic research; the other 90 percent focused on telephony, or technologies associated with communications. This subject has evolved from studies on how to transmit telephone calls across the United States (in 1907) to work on radio astronomy that led directly to our hearing noise from the "big bang" and understanding that the average temperature of the universe was 3 degrees above zero.

The genesis of the Bell Labs dates back to the nineteenth century and to the invention of the telephone by Alexander Graham Bell in 1875. In 1885 AT&T was established as the American Bell Telephone Company under the tutelage of Theodore N. Vail. In 1881 technical problems and demands for new services and capabilities led to the establishment of the Electrical Department and to the Mechanical Department (also called the Engineers Department) in 1884. They merged in 1902 to form an organization of some 200 people. Almost from its earliest days the research arm of the Bell world hired Ph.D.s. The first was William Jacques (Ph.D. from Johns Hopkins University in physics, 1879), and another came in 1885: Hammond Vinton Hayes (holding the second Ph.D. given by Harvard in physics). These and other scientists established at what would someday be called Bell Labs a tone of research and a sophisticated level of scientific endeavor. From the beginning to the present the management at Bell Labs has always been made up of scientists with considerable achievements to their credit.

In 1897 George Campbell came to work for the department to find a better way to carry a telephone call across the nation. The project launched a long line of efforts to develop signal transmission technologies and message switching equipment. These forced Bell Labs to study all aspects of electronics and solid-state physics, and thus ultimately to play an early and significant role in the development of computers, transistors, and semiconductors. Bell Labs also developed highly specialized techniques for packaging such technologies into usable and frequently cost-effective forms. AT&T was, for example, the first major institution to use Lee de Forest's repeater (invented in 1906) to amplify signals. His "audion" became the three-element vacuum tube amplified which in modified form was the heart of the first digital computers* of the 1940s. In the 1950s vacuum tubes gave way to a Bell invention, the transistor. New repeaters went into use in telephone communications on a regular basis as early as 1913. During 1914 about 550 engineers and scientists were working for the phone company in New York. The staff grew to 3,000 by early 1924.

Effective January 1, 1925, the name of this organization was changed to Bell Telephone Laboratories. Bell Labs, as it was and continues to be known, had formal responsibility for developing all technologies which Western Electric needed for the manufacture of telephone equipment. All the technologies used to switch calls from one city to another, for example, originated with work done at Bell Labs. Again, numbers suggest the growth and importance of this mission, a direction that has remained essentially intact to the present. Using 1981 as an example, that year researchers there were issued 311 patents, delivered or published 5,725 papers, and garnished 89 awards. Between 1925, when it had 3,000 employees, and 1981, the headcount increased to 24,078. Of that number in 1981, 3,328 had Ph.D.s, another 5,753 master's, and 4,007 just the bachelor's degree—some 13,000 with college and university degrees, as many as populated a large American university. Most of these researchers were also under the age of forty, which suggested many more years of significant work on leading-edge projects. Bell Labs had reached the point where it operated twenty-one laboratories as well. Until the 1930s solid-state physics was hardly studied; after that time it grew, achieving a very important position by the 1950s and 1960s. Bell Labs' budget in 1981 was $1.63 billion.

Bell Labs' first important foray into what would become the world of computers dates back to the early 1900s. The century opened with the electromechanical relay as a principal device usable for switching telephone calls across a network which, by 1910, involved over 7.5 million telephones in the United States alone. The use of relay technologies within the telephone system meant that at Bell Labs enough people were together who understood such devices to take on complex research projects involving the use of electricity to carry data (voice and numbers). Their work led to efficient and cost-effective devices by the late 1920s. By the 1930s the development of such equipment involved extensive calculation of numbers, and desktop calculators simply could not do the job. A mathematician at Bell Labs, George Stibitz,** began to experiment with scrap relays in 1937 and by 1938 concluded that he could build a device to perform complex calculations.

Under the direction of S. B. Williams and with George Stibitz heavily involved, Bell Labs undertook to build what became known as relay computers. By spring 1939, Stibitz had such a device designed; it became operational on January 8, 1940, and was kept in use until 1949. It was called the Complex Number Calculator and could do normal mathematical functions such as add, subtract, multiply, and divide. It serviced more than one terminal. On September 11, 1940, it did work on a remote basis when Stibitz submitted tasks to it from Dartmouth College over a telephone line. The device returned responses to it over these same special lines—the first known case of remote job entry (RJE) in the history of data processing. One of those who witnessed the event, John Mauchly,** went on to help build the ENIAC* in the mid–1940s, the world's first electronic digital computer, while another, Norbert Weiner,** who invented the term *cybernetics* in the late 1940s, became one of the patron saints of artificial intelligence.*

When World War II began, Bell Labs' skills were immediately dedicated to national defense. During the war, for example, it conducted nearly half of all radar-related research in the United States. It also did work on devices to supersede Stibitz's first equipment. This involved the construction of a machine that could perform automatic aiming for antiaircraft guns. The military wanted a machine that aimed a gun as a gunner tracked an airplane by turning knobs. The result was the Model II Relay Calculator, also called the Relay Interpolator, because it initially did linear interpolations and smoothing on data that came from paper tape.* The device became operational in September 1943. This machine started the Model series, with Stibitz improving on the design and performance of the basic computer. The Model II had 493 relays and was about five feet high and four feet long. After World War II and until 1961 it operated for the U.S. Naval Research Laboratory.

A third relay computer that came from Stibitz was called the Model III. Design work began in 1942, and its purpose was to simulate the flight of a ballistic projector, hence its nickname "Ballistic Computer." It incorporated the technology of its immediate predecessor and yet was more powerful. In its final form in June 1944, when it first became operational, it had 1,300 relays and 35 multicontact relays (for use in the multiply/divide unit), and had grown to 15 feet in width yet still 5 feet high. In 1948 it was moved to the Army Field Forces Board at Fort Bliss, Texas, and ran until 1958. A Model IV was also built for the U.S. Navy; it was completed in March 1945 and operated until 1961.

Two copies of the Model V were constructed. The first was delivered in July 1946 and the second in February 1947. They were also known as the Bell Laboratories General Purpose Relay Calculators, and because each could be split in half to work on separate problems, it represented one of the first computers that had what later would be called "partitions." Scientists at Bell Labs made one final computer in the series called the Model VI, which was similar to the V and was used only at the Murray Hill complex to do jobs originally performed by the Complex Number Calculator. It could manipulate 10-digit instead of 7-digit numbers, contained only one arithmetic unit, and handled up to 200 hardwired subroutines. Although Bell Labs experimented with other computers, the next major event there to influence data processing and the world of electronics in general was the invention of the transistor, one of the most important technological developments of midcentury.

Most histories of the transistor begin with events at Bell Labs in the 1940s. In fact, scientists there had struggled with the problem of how to manage electrical currents for a much longer time, as well as with the issue of how to amplify them. As early as the first half of the 1930s, they had concluded that the technological limits of vacuum tubes would soon be reached. Tubes had been used since around 1915 and were improved over time, but they remained large, expensive, and insufficiently reliable. To attack the problem aggressively, in the middle of the 1930s Bell Labs launched a program in solid-state physics in hopes of finding a better way. That effort led directly to the invention of the transistor. The first major step came in 1940 when Russel Ohl, part of the radio group at Bell, worked on the problem of how to detect short-wave radio signals as opposed to the more conventional vacuum-tube circuit transmittals that were longer. Already in 1939, after examining over 100 crystalline materials, he decided to study silicon, and the following year he stumbled across the photovoltaic effect with silicon. In effect, when light hit the silicon just right, an electrical current was produced. One of the future inventors of the transistor, Walter Brattain, was impressed with this discovery, but nothing came of it because World War II interfered. One byproduct of Ohl's discovery came in 1954 when three scientists at Bell Labs picked up where he left off and developed the first solar battery.

This background is important because by 1940, and especially during the early years of that decade, Bell Labs had acquired considerable experience with silicon, a semiconductor of electricity. Bell's experience with amplification of electronics, as a means of improving telephone and radio communications, put it in one of the best positions to appreciate the potential of silicon while highly motivated to improve existing technologies. Bell Labs had also become a very large operation by 1945: it had some 8,000 employees and thus the resources with which to push forward with new lines of research.

William Shockley,** one of the inventors of the transistor, had been at Bell Labs since 1936. Like Walter Brattain who participated in similar research, he spent the war years working on submarine detection research. In 1944 they were back at Bell Labs, and in July 1945 its executive vice-president, Mervin J. Kelly, instituted a major reorganization. One result of this exercise was the establishment of a major program to do fundamental research on semiconductors and solid-state physics under Shockley's direction. He in turn brought John Bardeen into the group along with Gerald Pearson who had been conducting experiments using semiconductors for a decade. Although the history of the transistor and consequently of the chip* may be found elsewhere in this dictionary, what is important to remember here is that it barely took any time before they developed the transistor, consisting of nonmoving parts in which silicon amplified electrical impulses. This event took place in 1948, so rapidly that it caught the lab and AT&T by surprise. But everyone appreciated the fact that it meant the end of the vacuum tube and the dawn of a new technology.

By the early 1950s Bell Labs developed better devices and efficient ways to manufacture them. It licensed other firms to make such units, and, by the end of the 1950s, computers were being built with these components. Transistors were less expensive, more efficient, and faster than tubes and thus made possible the existence of larger computers with more capacity and greater reliability. The impact on electronics in general was impressive, driving down costs and increasing efficiencies. It made practical the microelectronics essential for modern weapons, especially rockets, and gave scientists the ability to make spaceships and portable radios. For discovering the silicon effect, Bardeen, Brattain, and Shockley shared the Nobel Prize for Physics in 1956. Bardeen won a second Nobel Prize in Physics in 1972 for work done on superconductivity. No other person has ever won the same prize twice in the same field. Shockley went on to form his own companies to produce chips in the 1950s and 1960s. Brattain continued to do research.

The 1950s and 1960s at Bell Labs were busy decades for computer-related research. The initiative on development shifted westward to Texas Instruments† and to the Silicon Valley, leading to the development of the computer chip in the late 1950s and its wide use by the late 1960s. Work at Bell Labs focused on topics relevant to the telephone business or on pure research that, though dramatic, was not necessarily of direct importance to data processing. However,

Bell participated in work on integrated circuits which relied on the use of silicon chips.

In 1954 Carl J. Frosch and L. Derick, both of Bell Labs, completed work which culminated in the integrated circuit becoming a practical reality. The idea of putting an integrated circuit on a silicon chip was the work of two men outside of Bell Labs—Shockley and Jack S. Kilby** (of Texas Instruments)—in the late 1950s. Frosch and Derick found out how to make an integrated circuit as a result of their attempt to impose impurities on the surface of a silicon wafer to make transistors. They found that heated gas with impurities exposed to the surface of silicon formed a layer on it. This protective surface was both electrically and chemically neutral and, in effect, prevented other elements from influencing the behavior of the silicon. Frosch and Derick then found that they could precisely etch paths through this surface by using hydrofluoric acid, putting impurities in these paths, and thereby dictating how electrical currents would flow through the silicon. Although the above explanation is almost too simplistic, one can appreciate how an integrated circuit was made using silicon. Circuits of the 1980s have many layers of such coatings, forming a sort of "apartment house" of paths for electrons to pass through on a single chip. Germanium, the other popular substance studied in the 1950s as a possible basis for a chip, did not react the same way to this kind of surface coating, making silicon the winner by the end of the 1950s. It became the most widely used substance for chips in the 1960s.

Although the process that led to the manufacture of chips was far more complicated than one might be led to believe here, the historical point is that Bell Labs worked on the problem of how to make circuits. Bell Labs took the issue seriously enough to establish an Integrated Circuit Design Capability Laboratory called the Blue Zoo at Murray Hill (named after the azure surgical gowns worn there). Throughout the 1960s and 1970s work on circuits continued. The importance of such research became apparent only after AT&T was forced to compete in the telecommunications arena in the 1980s and decided to produce computers to rival those of companies like International Business Machines Corporation (IBM).† That new mission put increased pressure on Bell Labs to develop computer products.

In 1962 the first Josephson junctions were made at Bell Labs. This might constitute a major historical event, but as of 1986, the technical and business cases for their use have not yet become strong enough. In effect, this technology calls for the use of superconductive materials that totally eliminates electrical resistance. This only happens at temperatures near absolute zero or −273.15°C. Brian Josephson, a student at Cambridge University at the time, got the idea for this technology from a Bell scientist, Philip Anderson. This technology promises very high speeds at low costs. The issue of costs has not yet been resolved by computer manufacturers. Anderson went on to other projects, continuing Bell Labs' tradition of excellence and pioneering work in solid-state physics, winning a Nobel Prize for Physics in 1977 for his research in quantum mechanics. Most writers on the history of data processing's technology have dismissed Josephson technologies as switches whose time has not come. This may be so, but such technology has been used for complex measuring equipment, particularly within Bell Labs.

Bell Labs made yet another contribution to data processing in the early 1960s: the development of the UNIX operating system. Every large computer has a collection of software to manage the influx of application programs, data for those programs, and the use of peripheral equipment in a coordinated fashion. All of these household tasks are managed by an operating system. One of the most sophisticated systems to appear in the 1960s was UNIX, although it did not become widely available to users outside of Bell Labs until the 1970s. It was written by Ken Thompson and D. M. Ritchie. This particular operating system was one very large collection of integrated programs, as were most operating systems, and was modular. A series of small programs could be pulled together to make an operating system for any size computer, from a mini to a large mainframe; the code was also efficient and worked well. Thompson's and Ritchie's work represented some of the most sophisticated in the field during the 1960s and 1970s. Thompson was inducted into the National Academy of Engineering, while both he and Ritchie were given the A. M. Turing Award in October 1983 as recognition from the data processing industry for their work.

The promise of past successes and the needs and hopes of the present suggest that Bell Labs should continue its significant role. The promise looming on the horizon includes fiber optics (the use of light to deliver messages) discernible by computers. By 1984, for example, AT&T was investing some $100 million each year in this technology and in February 1983 had opened a 372-mile fiber-optic line between Washington, D.C., and the city of New York. This event portended of things to come, and since 1986, the Bell network has been converting to optics. This technology takes less space than cables and wires, is easier to install, and handles larger quantities of data. It is also proving less expensive. Optics may also someday eliminate many telephone poles, that ubiquitous visual feature of the twentieth-century landscape.

In assessing Bell Labs' impact on data processing, several themes become obvious. First, it produced some of the leading technologies of its day, and in some cases, as with the transistor, the most important. Second, quality workmanship appears as a way of life—quality in the scientists hired, quality in the projects undertaken, quality in the management and support for the entire enterprise from the Bell empire, particularly out of AT&T. It is an institution of enormous size and productivity when compared to American and European universities. In the 1980s its budget was equal to 10 to 15 percent of what the National Science Foundation spent. Yet for all its personnel and talents, its contributions have been no more numerous than what came out of the University of Pennsylvania in the 1940s (ENIAC), from Purdue University, or IBM. The difference, however, is that, decade after decade, Bell Labs has contributed to the technology that has made it possible for computers to switch messages around electronically. Universities, on the other hand, have made abundant contributions but not over such a sustained period. Only IBM has continued to produce new technologies and to convert them into usable items as frequently as Bell Labs, but IBM has not been around as long as Bell. Over the long haul, the most important work at Bell Labs has involved research in solid-state physics and their applications to electronics.

For further information, see: E. G. Andrews, "Telephone Switching and the Early Bell Laboratories Computers," *Annals of the History of Computing* 4, no. 1 (January 1982): 13–19; Bell Telephone Laboratories, *Facts About Bell Labs*, 12th ed. (Murray Hill, N.J.: Bell Telephone Laboratories, 1982); Jeremy Bernstein, *Three Degrees Above Zero: Bell Labs in the Information Age* (New York: Charles Scribner's Sons, 1984); Ernest Braun and Stuart Macdonald, *Revolution in Miniature: The History and Impact of Semiconductor Electronics* (New York: Cambridge University Press, 1978; ed., 1983); Allan G. Bromley, "Origins of Antiaircraft Analog Computers," *Annals of the History of Computing* 6, no. 2 (April 1984): 163–164; W. H. C. Higgins et al., "Defense Research at Bell Labs," ibid., 4, no. 3 (July 1982): 218–236; Lillian Hartmann Hoddeson, "The Roots of Solid-State Research at Bell Labs," *Physics Today* 30 (March 1977): 23–30, *Multidisciplinary Research in Mission-Oriented Laboratories: The Evolution of Bell Laboratories' Program in Basic Solid-State Physics Culminating in the Discovery of the Transistor, 1935–1948* (Urbana: University of Illinois Press, 1978), and "The Emergence of Basic Research in the Bell Telephone System, 1875–1915," *Technology and Culture* 22 (July 1981): 512–544; Bernard D. Holbrook and W. Stanley Brown, *A History of Computing Research at Bell Laboratories (1937–1975)*, Computing Science Technical Report No. 99 (Murray Hill, N.J.: Bell Telephone Laboratories, 1982); George R. Stibitz, "Early Computers," in N. Metropolis et al, eds., *A History of Computing in the Twentieth Century* (New York: Academic Press, 1980): 479–483, and "The Relay Computers at Bell Labs," *Datamation* 13, no. 4 (April 1967): 35–44 and no. 5 (May 1967): 45–49.

BLETCHLEY PARK. This was the site of secret British cipher activities during World War II. Although much of the detailed history of activities here are still shrouded in secrecy, it is known that computational projects were developed there to support British efforts at deciphering German codes. One of these projects involved the construction of the COLOSSUS.* Many of the scientists who worked at Bletchley Park went on to conduct research in the development of computers in the late 1940s and throughout the 1950s.

Bletchley Park was the British Post Office Research Station managed for the Government Code and Cipher School in Buckinghamshire. It had been an estate encompassing various buildings on its grounds in which different groups worked in secret, frequently not knowing what other groups in the park were doing. Apparently, the primary mission of this group was to break German codes. There were two groups. The first was generated by a machine called ENIGMA,* often described as the Enigma series. This device scrambled each letter of a message, was portable, and was widely used for routine, daily military communications. The second system, called the Geheimschreiber, was dedicated to top secret communications. The British built machines that broke the codes of both systems. The Geheimschreiber system required the construction of a computer called COLOSSUS. Mathematicians worked out methods for determining the best way to crack codes, while engineers built machines to carry out these procedures.

For further information, see: Patrick Beesly, *Intelligence: The Story of the Admiralty's Operational Intelligence Centre* (London: Hamish Hamilton, 1977); Anthony C. Brown, *Bodyguard of Lies* (New York: Harper & Row, 1975); I. J. Good, "Early Work on Computers at Bletchley," *Annals of the History of Computing* 1, no. 1 (July 1979): 38–48; Brian Johnson, *The Secret War* (London: British Broadcasting Corporation, 1978); Reginald Victor Jones, *The Wizard War: British Scientific Intelligence, 1939–1945* (New York: Coward, McCann & Geoghegan, 1978); David Kahn, *The Code Breakers* (New York: Macmillan, 1967); Ronald Lewin, *Ultra Goes to War* (London: Hutchinson, 1978); Frederick W. Winterbotham, *The Ultra Secret* (New York: Harper & Row, 1974).

BRITISH TABULATING MACHINE COMPANY (BTM). BTM is one of the oldest tabulating companies in Europe and has stayed in business by marketing computers. This company was established by Robert Percival Porter, an English-born journalist and a friend of Herman Hollerith,** inventor of the modern tabulating equipment. The two men had attempted to found a tabulating firm in Great Britain in 1894 with little success. Seven years later, Hollerith persuaded Porter to try again. In 1902 the idea was to have it partially capitalized by Hollerith's firm, with the rest of the money raised in England. The new company was to have access to Hollerith's patents. In 1904 the Tabulator Limited came into existence, and soon after a customer was found, the Woolwich Arsenal Ordnance Factory. In 1907 this small company became known as the British Tabulating Machine Company capitalized at £50,000, remaining tied first to Hollerith's company, and after he sold it, to International Business Machines Corporation (IBM)† until 1949. In 1949 IBM broke with the firm and established its own, called IBM United Kingdom. The new company competed against BTM.

BTM remained a small enterprise until it landed a large contract for tabulating equipment with the British government for use in taking the Census of 1911. During World War I equipment was leased to the British government for military purposes, much as happened in the United States. In the 1920s and 1930s tabulating equipment made its appearance in businesses for basic accounting applications in manufacturing, insurance, and government.

Following World War II, BTM became interested in computers as a logical extension of its information processing business. A machine built by A. D. Booth at Birkbeck College, London University, caught the firm's interest and was acquired. Called the Hollerith Electronic Computer (HEC), it was first shown to the public in 1953 at the Business Efficiency Exhibition. It was sold commercially as the BTM 1200 beginning in 1954. That year five were sold, and in 1956, a larger version, called the BTM 1201, was shipped with a 1,024 word drum storage. During this product's life, seventy were sold. BTM competed in the 1950s against IBM and J. Lyons & Company, competitors who often used

BTM's card punch equipment as peripherals on their systems. Punched card equipment remained the staple of the firm until the early 1960s when computers and other related products dominated sales. During the 1960s BTM had become a fully computerized enterprise.

A major change in the company took place in 1959 when it merged with the Powers-Samas Accounting Machines, Ltd. to form International Computers and Tabulators (ICT). In 1963 it bought out Ferranti Company's† computer interests. Earlier, in 1961, it had acquired General Electric Company Ltd. (GEC)'s British computer operations and the following year those belonging to Electrical and Musical Industries Ltd. (EMI). Thus, beginning in January 1963, ICT could claim that 25 percent of all installed computers in Britain had come from it. Ferranti had had a similar percentage. Other major vendors in Great Britain had the following shares of net worth of installed computers: English Electric Company (13 percent), Elliott Brothers (12 percent), National Cash Register Company (NCR)† (11 percent), Leo Computers Ltd. (7 percent), and others making up the remaining 7 percent of installed computers. In 1968 English Electric merged with ICT, creating International Computers Ltd. (ICL).

The history of these firms strongly suggests when computers made their impact on Great Britain. Sales were minor for computers in comparison to tabulating equipment for all companies. In 1957 computer sales jumped by 100 percent and, after 1962, soared. Statistics suggest that the use of computers became widespread in British society in 1966. One of the more important machines from the 1960s was the 1900 series which remained in the product line until the end of the 1970s. By 1968 alone, over 1,000 copies had been installed. It was followed in 1974 with the ICL 2900 series.

For further information, see: Simon Lavington, *Early British Computers* (Bedford, Mass.: Digital Press, 1980).

BTM. See BRITISH TABULATING MACHINE COMPANY

BULL COMPANY. The Bull Company was a leading French computer manufacturing firm in the 1950s. It was established in 1922 by a Norwegian engineer, Frederick Bull (1882–1925), and by a fellow scientist, Knut Kruesen, for the purpose of manufacturing and selling punched card equipment. The company sold twenty such devices by the end of 1925 in Europe. In 1929 H. W. Egli, along with other Swiss investors, acquired the patents on this equipment and then established the Egli-Bull Company. In 1932 a French group called Caillies bought the firm and renamed it the Société des Machines Bull, and since then it has usually been called Machines Bull. During the 1920s and 1930s Bull managed to sell tabulating equipment typical of the period, but during World War II the firm barely survived. Following the war, the company enjoyed a revival that made it one of the most important computer manufacturers in Europe.

During the 1950s the Bull Company became interested in computers and in

1958 brought out its first machine which became known as the Gamma Extension Tambour, or simply the Gamma ET. It relied on delay lines for main memory and a drum for additional storage. Throughout the early 1950s the company had marketed various calculators that gave it the experience it needed to move into computers more rapidly. The inventor of the word *informatique*, Philippe Dreyfus, was associated with the firm. The first computer built by the company, the Gamma 2, was introduced to the public in 1951. This calculator used delay-line memory and germanium diodes for logical circuits. Peripheral equipment consisted of card punch equipment. Gamma 3 appeared in 1952, and some 1,000 were sold. Other versions with enhanced memory capabilities appeared, but all were fundamentally calculators rather than stored program computers.

Late in the period of first-generation computers, the Gamma ET relied heavily on the Gamma 3 but with enlarged memory. Bull later introduced the Gamma 60 which competed against the IBM 7030 (STRETCH)* and the LARC* at the end of the 1950s and was the company's major product of the early 1960s. By the end of its life (in the mid–1960s) some one dozen had been sold and installed. After the early 1960s, other computer firms took precedence in the French market, thereby decreasing the significance of the early work done at Bull.

For further information, see: René Moreau, *The Computer Comes of Age: The People, the Hardware, and the Software* (Cambridge, Mass.: MIT Press, 1984).

BURROUGHS CORPORATION. This company was a leading supplier of office machines in the first four decades of the twentieth century and, like many other suppliers of office equipment, moved into the data processing industry after World War II. The company experienced a period as the leading supplier of office-related machines (such as adding and tabulating equipment) in the 1920s and a period of eclipse during the 1930s and 1940s to emerge from World War II with the interest and technical capability to move into electronics. While behind some other companies during the 1950s and 1960s in the development of computers, it nonetheless emerged in the late 1970s as the second largest supplier of data products in the computer industry. It had made the transition from a broad-based supplier of mechanical adding machinery, primarily to banks, to a fully integrated supplier of computers, peripheral equipment, and software. Its history largely mirrors that of other companies such as International Business Machines Corporation (IBM),† Honeywell,† and other members of the data processing industry.

The company was formed by William S. Burroughs (1855–1898), a one-time bank clerk who, like his peers, calculated figures by hand at the cost of very long work-days. His career as a business machine supplier grew out of his desire to automate an accounting clerk's function. Although many other individuals were developing mechanical aids to calculation in the nineteenth century, his work ultimately led to the creation of a company that became the leader in its marketplace. He thus built a practical adding machine but, unlike most of his peers, he had the necessary business skills to manufacture and market his products. By the early twentieth century the Burroughs Adding and Listing Machine was one of the most popular business devices available. It enabled clerks to add numbers many times faster and more accurately than clerks calculating by hand. Indeed, a less well-paid clerk could do more work with such an adding machine than a better paid accountant working by hand. These machines made it possible for banks to keep records current, the insurance industry to develop and maintain its necessary charts and records, and the railroad industry to monitor its vast operations.

Burroughs, like many others who dabbled in office machinery, was no stranger to engines. He was the son of a mechanic, and he himself worked with machines. His first employment was as a clerk in a bank where long hours of work adding numbers led to a decline in his health, forcing him at the age of twenty-four to leave his job. He next took up his father's profession and at the same time was determined to invent a device that banks could use for calculating and tabulating numbers. By 1880 he had established his own company which by 1885 was called the Arithmometer Company and produced various office machines. The development of these machines continued throughout the 1880s and 1890s. In 1889, Burroughs' company built fifty boxey devices for the purpose of adding that proved to be large, heavy, and unpopular. As a result, the company began to give more consideration to smaller units that would be easier to use. A number of new products were widely introduced in both the United States and Europe by the start of World War I. They generally were lightweight (although made of black painted metal) but could be moved and set up on a desk. They were both reliable and fast. When these products gained acceptance, the story that circulated was that Burroughs threw his original fifty products out the window, smashing them on the pavement below. The company had learned that its focus should be on the characteristics a product needed rather than on the development of a machine for the sake of doing a particular function. Thus, it established the principle that technological function had to be married to customer demands, a principle accepted by all office machine companies early on.

This concept influenced the development of computers in later decades. Technology in itself would not be the only consideration; particular emphasis would be placed on ease of use and the applications for such devices. This emphasis led the Burroughs Machine Company (as it was known before World War II) to introduce a variety of new products, especially after World War I. During the 1920s the company introduced the Burroughs Class 16 machine which allowed a user to accumulate numbers in several registers as opposed to adding numbers in simple columns. Leslie J. Comrie,** the famous British astronomer and an authority on the use of business machines, used this device almost as a difference engine to create and audit nautical tables, thereby dramatically illustrating how the scientific community could use accounting equipment. He called the Class 16, along with the National-Ellis 3000, "modern Babbage Machines" that did "what the Babbage difference engine* was intended to do."

During these years the Burroughs organization was managed by a conservative team of executives who were well versed in mechanical office machinery. They knew adding machines and cash registers, and had made sales primarily to banks, railroads, and insurance companies. In addition to selling a reliable line of equipment, they had expanded the business into a major, and perhaps dominant, force in the office equipment market by sound acquisitions. Thus, for example, in 1921 they purchased the Moon-Hopkins Billing Machine Company to enhance Burroughs's role as the strongest supplier of office machinery. By the early 1920s sales were international and factories existed in the United States, Europe, and Latin America. Profits were high: $8.3 million on sales of $32 million in 1928, for example. Revenues for that year were nearly twice IBM's $19.7. IBM, its primary competition in the office arena, made profits of only $5.3 million that year. Another major competitor, National Cash Register (NCR),† had sales of $49 million but profits of only $7.8 million in 1928. Clearly, Burroughs was the most profitable company in the industry.

Although the Depression of the 1930s dampened sales and profits, at the start of World War II Burroughs was still a broad-based supplier of office equipment which included adding machines, bookkeeping devices, accounting units, and cash registers. Some of its leading competitors would subsequently be rivals in the data processing industry, including IBM, NCR, and Honeywell. Like other companies with considerable machining expertise, during World War II the giant supplier of adding machines built war-related equipment for the U.S. government. Its most famous contribution to the Allied war effort was the manufacture of the Norden bombsight. The company's full attention was focused on government contracts; almost all of its output went to Washington.

The war dramatically influenced the future directions of the company. It exposed its managers to government contracts and to the possibilities of business with the military establishment. Burroughs became more sensitive to military applications which would prove lucrative in the 1950s. A most important implication for the development of computers at Burroughs was the company's awareness of the possibilities of electronics. The bombsight, although dramatic evidence of the new technologies being used at the company, was only one of many projects. As managers became more aware of electronics and as researchers within the company became attuned to its application to possible products, the traditional focus on mechanical office equipment ended. Thus, by the end of World War II, both senior management and newer arrivals to the company recognized that the future would require products based on electronic technologies.

This fundamental shift in focus coincided with the more immediate necessity in 1945 of the shift in clientele from the government to Burroughs's more traditional commercial buyers. The shift came as the company recognized that new products would have to be developed that did more than the mechanical products of earlier decades. Revenues and profits suggested the tenuousness of the new situation. In 1946 profits were $1.9 million on revenues of about $100 million. That same year $1 million was budgeted for research, and by the end of the 1940s this figure had risen to over $3 million per year. John Coleman, president of the company immediately after World War II, is credited with enabling Burroughs to enter the computer age. He also encouraged the hiring of scientists who were university trained and created the opportunity for them to move into the ranks of management. Thus, by the mid–1950s a large number of key executives did not have an accounting machine background of the typical executive.

The company's first important data processing product came in 1953 in the form of static magnetic memory to be used as part of the ENIAC* computer owned by the U.S. Army. This additional memory allowed the ENIAC to store six times more data in memory. At the same time Burroughs developed a computer called the Unitized Digital Electronic Computer (UDEC) for experimental purposes. These two developments staked out an important position for Burroughs in the newly developing computer field. This company, like its peers, had a difficult time finding enough computer specialists to hire and thus contributed the UDEC to Wayne State University for the purpose of training badly needed specialists. Other models of this machine were worked on, and in 1955 Burroughs introduced the UDEC II. The machine was touted as a computer designed to solve complicated mathematical problems in such business areas as production scheduling, cost analysis, inventory control, product design, and forecasting. Like those machines made by such companies as IBM, General Electric (GE),† and Honeywell, it reflected an early and growing concern to apply electronics to business applications.

Burrough's first widely available commercial computer was the E–101, which appeared in 1954. It was desk-sized and cost about $35,000, and was built and priced to be sold in large numbers. Burroughs hoped to market the device to its customer set, many of whom were in the banking industry, but the product was a failure. It had too few functions for either the scientific or business community. Despite its failure, the company continued working on various defense projects involving computers. Much of the research and experience gained with these contracts would be applied in commercial products later in the 1950s. During this decade, the company made quiet inroads into the new field of data processing, one that required computers, peripheral equipment, software, new uses for business machines, and equally creative ways of marketing. As with many other companies in data processing, Burroughs's most important customer for its early devices was the federal government. In 1956 large-scale computers were built as part of the SAGE* project. That contract alone led to almost $40 million of revenue by the end of 1957. The year before, the company purchased Electrodata, a small firm that developed computerized technology, as a means of furthering its expertise in electronics. Electrodata made the DATATRON 220, a vacuum tube computer which appeared in 1957. At the time of its introduction, however, it was technologically inferior to products appearing in the marketplace such as the IBM 7070 and 1401.*

Despite this setback, Burroughs continued research and development of computers for the next several years, but was cautious in introducing new products. It reentered the market in full force in the early 1960s with a variety of products. By the end of 1963 these products had contributed substantially to the company's data processing revenues in the United States of some $42 million. These products were still primarily based on electromechanical technologies that were rapidly becoming obsolete.

IBM rocked the industry in April 1964 with its introduction of a family of computers known as the S/360.* This set of devices, complete with peripherals and software, represented a major leap forward in technology and a significant drop in the cost of computing. In short, it was nothing less than the introduction of a new generation of computers. The giants of the young data processing industry responded with a host of products in the months and years to come. One was Burroughs which in August of that year introduced the B5500, the start of a family of computers called the 500. The data processing industry had reached the point where it was introducing a variety of computers of differing sizes but compatible with each other. That is, programs written for one could be run on other, larger machines of the same family. Until then, when a user needed a bigger computer, all existing programs had to be converted to operate in a newer device at enormous cost. In fact, some conversions cost more than the expensive computers themselves. And with the base of application programs increasing dramatically during the 1960s, the pressure to minimize conversions increased.

Burroughs added more members to the family during the mid–1960s. Joining the 500 family were the B6500, B2500, B3500, B5500, and the never shipped B8500. These computers varied in size, allowing the company to market to a broad set of customers, from medium-sized accounts to the largest banks and insurance companies.

With the variety of products that came out of Burroughs in the late 1950s and during the early years of the 1960s, the impression was that of a large product development group which seemed more massive than it really was. From the 1950s on, the company had small clusters of people working on products. For example, its software scientists in the 1950s usually numbered about twenty-five, headed by William R. Lonergen in the late 1950s. Within the company he is frequently credited with convincing the old-line Burroughs executives to change their views on computing from mechanical devices to high-speed software-driven computing products. Yet these old-line executives, while convinced and pushed into data processing, never lost their sense of what customers wanted. Thus, the focus at Burroughs was less on technological superiority (for example, in the construction of ever more powerful smaller computers) and more on the ease of use of such equipment for their customers. Long influenced by what many surveys throughout the first five decades of the century indicated to be the loyalist customer set in the business machines marketplace, the company focused on what their customers could use. This was particularly true in the banking community where the company was most entrenched.

Important software developments came out of Burroughs during these years. The operating system for the B5000 permitted job scheduling and allocation of memory and peripherals through software. The ability to run more than one program or job at the same time, today called multiprogramming, was a Burroughs first. It was one of the first, if not the initial, organization to develop virtual memory which allowed peripheral storage devices to store data that otherwise would have had to be in computer memory before a specific program could be run. These developments made it possible for computers to process more data than before and to do so easier. It meant that computer operators had to know less about computer technology than before to run processors. Yet despite these innovations and the company's concerns for the ease of use of such computers for customers, they did not sell well.

In the early 1960s the company's executives were disappointed with the response to their products. There were indicators of what happened. For one thing, the primary language of its new family of computers was ALGOL.* The current trends in programming favored FORTRAN,* Assembler, and, for commercial applications especially, COBOL.* The initial machines operated too slowly (a major complaint with the B5000), although this problem was fixed with subsequent machines. The change in speeds did help pick up sales. The company also improved its situation by building a sales force trained in selling computers and by relying less on the traditional Burroughs salesmen steeped in the adding machine heritage.

The company faced up to its problems by cutting back expenses, reallocating budgets, and putting more technologically experienced managers into positions of authority. These actions increased the company's well-being in the late 1960s. Its stock rose in value which, in turn, made more funds available for research and product development. In 1968 data processing products finally showed a profit at Burroughs. By the mid–1970s their data processing sales hovered between second and third to IBM's. The cost of products declined as the heavy investments in research paid off. As with other data processing companies, the price of computer memory dropped while the use of more sophisticated operating systems and hardware increased, making its products attractive at a time when the cost of programming was rising.

Much of the credit for the rebirth of Burroughs during the late 1960s and early 1970s belongs to Ray W. Macdonald. In 1966 he headed up marketing and subsequently went on to run the company, retiring in 1977. Although accused of never delegating authority and responsibility and of being autocratic, he had a clear vision about how to market computer technology. It was he, for example, who decided to produce faster running members of the 500 family and to develop a marketing force capable of selling computers.

Revenues during the early to mid–1960s remained weak. In 1964 worldwide revenues were $392 million in comparison with RCA's $2 billion, Honeywell's over $600 million, Sperry Rand's† $1.3 billion, and IBM's $3 billion. Macdonald increased the productivity of the sales force while driving down manufacturing costs. Revenues went from the low of $392 million in 1964 to $759 million in 1969. Data processing's contribution from the U.S. market went from $61 million to $260 million. Profits jumped by about 500 percent, and by 1975 revenues had grown fourfold over the previous ten years to $1.5 billion. Personnel between 1964 and the end of 1975 grew from 34,000 to 51,500. The company entered the second half of the 1970s with fifty-four plants and two under construction.

Thus, the history of Burroughs in the 1970s was one of growth and renewed vitality. It had a broad set of products that included computers, peripherals, software, and more traditional office machinery. It had a large manufacturing capability, a sophisticated research and development organization, and management experienced in marketing within the data processing industry. Throughout the 1970s and into the 1980s, like many leading computer-related companies, Burroughs continued to introduce new technological innovations. When IBM announced its S/370* family in October 1970, Burroughs reacted with the introduction of the 700 family in early 1971. This announcement incorporated the use of monolithic integrated circuitry, new peripherals (particularly important were its disk drives to compete with IBM's 3330s), and software. Additional members of the family followed: B 4700 in 1971, B 1700 in 1972, and the L series of microcomputers in the early 1970s. The L series had the capability of being remote data communications equipment that could send information back and forth to larger central site computers. They could also operate independently as stand-alone processors. The L series reflected the demand at that time for networks of computers of various sizes within an organization, which by the end of the 1970s was a common use of computing in most companies of any significant size.

The introduction of new, more competitive equipment allowed Burroughs to grow, despite momentary declines which it along with many other computer companies experienced in 1971 owing to a U.S. recession. In 1972 revenues exceeded $1 billion, whereas data processing products contributed the fast-growing amounts during the 1970s. For example, in 1973 equipment orders generated a growth in revenue in one year alone of 27 percent. By the end of 1979, revenues from data processing products had reached $1.2 billion. By this time Burroughs had brought out the 900 family of computers along with additional peripherals and software. Its products ranged from large computers down to microsystems. The 900 family especially was a major event for the company. First introduced in 1979, it consisted of the B 90, B 1900, B 3900, and B 6900. Additional enhancements came over the next several years, allowing the company to offer products in the intermediate and upper ranges to market against, for example, IBM, which was still its prime competitor. It entered the 1980s as the second largest supplier of data processing products after IBM in the industry. In 1986 it merged with Sperry to form Unisys.

For further information, see: L. J. Comrie, *Modern Babbage Machines. Bulletin* (London: Office Machinery Users Association, Ltd., 1932); Franklin M. Fisher et al., *IBM and the U.S. Data Processing Industry: An Economic History* (New York: Praeger Publishers, 1983); Katharine D. Fishman, *The Computer Establishment* (New York: Harper & Row, 1981); B. Morgan, *Total to Date: The Evolution of the Adding Machine; The Story of Burroughs* (London: Burroughs Machines, 1953); M. H. Weik, *A Survey of Domestic Electronic Digital Computing Systems* (Aberdeen Proving Ground, Md.: Ballistics Research Laboratory, 1955).

C

CBI. See CHARLES BABBAGE INSTITUTE FOR THE HISTORY OF INFORMATION PROCESSING

CDC. See CONTROL DATA CORPORATION

CHARLES BABBAGE INSTITUTE FOR THE HISTORY OF INFORMATION PROCESSING (CBI). This organization was named after the nineteenth-century inventor of analytical engines, Charles Babbage.** It was established in 1977 to encourage the study of the history of data processing. In practice it became a source of information on resources available for research and established an archive for appropriate materials which included company records and files of individual scientists, and has a growing collection of publications. It is the single most important organization dedicated to the study of the information revolution.

From its earliest days CBI initiated research and conducted seminars. It aggressively promoted an awareness of the subject, primarily within the large data processing community. Because of the way it was organized and supported, it received help and endorsements from historians, archivists, academicians, scientists, data processing professionals, business managers, and government officials in many countries, although its strongest base as of 1986 remains within Great Britain, Canada, and the United States. Its board of directors has twenty-five members drawn from industry and academe, and includes computer scientists and historians of science, as well as industrial leaders. The board's first meeting was held at the Smithsonian Institution's Museum of History and Technology in Washington, D.C., on January 30, 1979.

The moving force behind its creation in 1977 was Erwin Tomash. Later, he went on to promote the republication of books critical to the early history of data processing. In 1978 Paul Armer took over the post of executive secretary for CBI. A major step in CBI's history was taken on June 21, 1979, when Tomash and Albert S. Hoagland, then president of the American Federation of Information Processing Societies (AFIPS),† entered into an agreement whereby AFIPS later became a major supporter of CBI. This arrangement gave CBI de facto recognition as a serious organization within the data processing community.

CBI established its first headquarters in Palo Alto, California, in April 1978, and moved to its permanent home in the fall of 1980 at the University of Minnesota. There it established an archive and library, and began to build a staff. Each year CBI offered a fellowship to a graduate student in the area of data processing. It also launched an aggressive oral history program that initially focused on interviews of scientists and engineers who built early computers. CBI has also sought to build its information base on institutional histories of organizations and companies critical to the history of data processing. CBI publishes a newsletter on its activities. Its current address is CBI, University of Minnesota, 104 Walter Library, 117 Pleasant Street, SE, Minneapolis, Minnesota 55455; telephone number (612) 376–9336.

For further information, see: CBI, *The Charles Babbage Institute Newsletter* (1979–).

COMMODORE INTERNATIONAL, INC. This firm specialized in the manufacture of small desktop computers during the early 1980s and was one of the most important companies in that market. Along with Apple Computer† and International Business Machines Corporation (IBM),† it enjoyed considerable success as a supplier of microcomputers.

The company was the brainchild of Jack Tramiel who dominated the firm. Tramiel was born in 1927 in Lodz, Poland, and during the last year of World War II was imprisoned at Auschwitz and at Bergen-Belsen concentration camps. In 1947 he emigrated to the United States where he drove a taxicab in New York City. After a stint in the U.S. Army, he and a friend, Manny Kapp, opened a typewriter repair business in the Bronx in 1954 which they called Commodore Portable Typewriter. The business was moved to Toronto in 1956 and became known as the Commodore Business Machines Company. Tramiel began selling adding machines and a Czech typewriter. During the late 1960s he stopped offering adding machines and in the early 1970s turned his attention to hand-held calculators. In May 1974 Commodore went public, and its stock was listed on the American Stock Exchange. The company had thrived, but the following year the price of computer chips* dropped, causing the prices of hand-held calculators to plummet from approximately $100 each to nearly $10. Commodore, trapped with a large inventory of the more expensive units, lost $5 million on sales of $50 million that year. To prevent such a swing in earnings from recurring, Tramiel acquired MOS Technology, located in Norristown, Pennsylvania, one of Commodore's suppliers of chips.

An engineer at MOS, Chuck Paddle, interested Tramiel in the idea of making desktop computers. In January 1977 Commodore announced its first product, named PET. It proved marketable and was followed by larger models, specifically the VIC–20 and the most successful of all, the Commodore 64. The company sought out the low end of the microcomputer market, leaving to Apple, Radio Shack, and later, IBM, the upper and most expensive end of the market. The lower end had been defined in the late 1970s as consisting of those products selling for less than $500 each, which usually meant that the home user or students and some scientists were the primary customers. The company did well against other low-end marketing firms such as Texas Instruments† and Atari,† eventually driving them both out of the market.

Yet all was not well for the firm: Tramiel lacked both the interest and the skill necessary to build an organization. High turnover in personnel, particularly in management and sales, made circumstances at the company unstable. Nonetheless, Commodore enjoyed one of the fastest growth records of any microcomputer company of the early 1980s. Although the reasons will be debated for some time, it appears that on the one hand the company's owner and dominant force had good marketing instincts while on the other, the products were very popular among users. In 1982 alone the Commodore 64 was extremely popular, with over 1 million copies sold; it was available in Europe and in the United States. In 1983 a similar number were bought just in the last three months of the company's fiscal year. That was the year Texas Instruments and Mattel left the market.

The company's record for revenues was impressive. Sales in 1977 had reached $50 million with profits of $3.4 million, which meant a good recovery from the problems faced in 1975. By 1984, sales exceeded $1 billion, generating profits in excess of $100 million. The value of the company's stock grew sixtyfold between 1977 and the end of 1983. Tramiel left Commodore in 1983 and by the fall emerged as the chief executive officer at Atari,† Commodore's arch rival. After he persuaded various employees at Commodore to join Atari, he was promptly sued. The next president of Commodore was Marshall F. Smith who had a background in accounting and had been a controller at U.S. Steel.

For further information, see: Robert Levering et al., *The Computer Entrepreneurs: Who's Making It Big and How in America's Upstart Industry* (New York: New American Library, 1984).

COMPUTER RESEARCH CORPORATION (CRC). CRC, one of the earliest companies in the data processing industry, became thc early nucleus of the computer sector of the National Cash Register Company (NCR)† in the 1950s. CRC was an outgrowth of Northrop Aviation,† home of considerable activity with computers in the late 1940s for an American company.

CRC was founded in 1950 by five electrical engineers with specialties in missile-guidance systems. Their initial intent was to construct small computers for the U.S. military community. CRC built one of the first medium-priced, general-purpose processors. By 1952 it claimed it had three computers in its product line: one for use with applications in engineering, another for scientists, and the third for business. In reality, it was in the business of supplying specialized computers to the military. In 1953 NCR acquired the company for $1 million and, by the mid–1950s, invested an additional $4-$5 million in the acquisition. NCR was prepared to do this because it pinned high hopes on CRC as its vehicle for entering the computer business, using the fast means of acquiring expertise to do so. The first computer product from NCR was the CRC 102D, a scientific application processor. Throughout the 1950s, NCR made modest introductions of new products, the result of CRC's expertise. Yet in the final analysis, NCR focused primarily on its traditional markets. Data processing revenues remained low and, in some years during the 1950s, dipped below previous levels.

In its day CRC was a forward, high-technology firm and, like many others of the late 1940s or early 1950s, aimed its services at the U.S. military community. These companies, including the Eckert-Mauchly Computer Company and Engineering Research Associates (ERA),† were small but competent technically, and all serviced the federal government. The majority were acquired in the 1950s or 1960s by larger electronics and office supply firms and thus lost their individual corporate identity. At NCR, for example, CRC became the NCR Electronics Division. In that capacity, it installed thirty copies of the 102, one of the first such devices to rely on germanium diodes rather than vacuum tubes. That switch in technology contributed to its lower price when compared to older, comparably sized machines.

For further information, see: John Desmond, "Cash Registers to Mainframes: The Legend of NCR," *Computerworld*, March 18, 1985, pp. 69, 73; Franklin M. Fisher et al., *IBM and the U.S. Data Processing Industry: An Economic History* (New York: Praeger Publishers, 1983).

COMPUTER SCIENCES CORPORATION (CSC). This was one of the first companies formed to write software under contract either for end users or other firms selling software and hardware as its primary business. Formed in the late 1950s, it survived over a quarter of a century in an industry in which the life span of software companies averaged less than a decade.

The company was launched on April 16, 1959, by two data processing managers and technocrats: Fletcher Jones, head of the data center for North American Aviation in Columbus, Ohio; and Roy Nutt, an employee of the United Aircraft Research Computation Laboratory. As a result of being in the aerospace industry, which unlike other economic groups had at an early date relied heavily on data processing, Nutt and Jones had become aware of the growing demand and importance of software. They had also been founding members of SHARE,† a users' group which, in the 1950s, consisted of sixteen to twenty aerospace companies that shared information about software. CSC's first contract was negotiated with Honeywell† to write a compiler in 1959. At the time, CSC was capitalized at $100.

CRC's first customers were manufacturing firms in the business of building computers and some government agencies. By the mid–1960s federal accounts had become very important to CSC. In 1965 the firm acquired two small operations from International Telephone and Telegraph (ITT) called Communications Systems, Inc. and Intelcom, giving CSC expertise in communications and access to U.S. government accounts. Some important early contracts were signed with the U.S. Atomic Energy Commission and with the NASA Marshall Space Flight Center. CSC wrote software and integrated systems, that is, it took the hardware of various vendors, wrote software for them such that all these components worked together, and sometimes operated them. From an historical perspective, one of the most important pieces of software which CSC developed was the early compiler for Honeywell. It was called the Fully Automatic Compiling Technique (FACT),* completed during 1959, and was a business-oriented language. That compiler (language) became a precursor to the industry's most widely used commercial programming language,* COBOL.* Military contracts were typically also for languages and some applications. An example from this period was an algebraic compiler which CSC developed for the U.S. military that ran on a AN/FSQ computer.

By the mid–1960s, when the company was prospering, there were between forty and fifty independent suppliers of programming services. In addition, an estimated 30 to 50 percent of all the systems software developed for the manufacturers of third-generation computers (circa 1964–1970) acquired these programs from software programming houses. The introduction and acceptance of the IBM S/360* encouraged many firms to form highly profitable software ventures; these firms required no capital other than the ability to program. CSC did very well in this environment and, by the end of 1969, had annual revenues of $67.2 million. It was a boom business: by the same year over 2,800 software houses were in operation.

CSC's early expertise was primarily in the area of systems software with programs that operated and managed computers and peripheral equipment. This company wrote more programs of this kind in the early 1960s than any other vendor. In addition to the FACT compiler, CSC also wrote the LARC* scientific compiler for Univac, followed by the development of the entire operating system for the UNIVAC* 1107 computer. By the early 1970s CSC had written systems programs for about fifty different types of computers. It also developed application programs including a tax preparation package, another to handle ticket services, and a dozen software tools that ran on S/360s for general ledger, accounts receivable, payroll, personnel, loans, and so on, these were all sold directly by CSC to end users. By the start of the 1970s, the company also marketed its own communications network manager called INFONET, relying on a UNIVAC 1108 to drive the system. Thus, during the 1960s a great deal of work had been done at CSC; revenues went from $5.7 million in 1964 to $17.8 million the next year and to $82 million in 1970.

In 1972 the General Services Administration began to standardize nonmilitary communications for remote data processing with INFONET. Additional contracts with military and civilian agencies meant that, by the end of the decade, CSC had developed some 100 different programming languages. Another important contract demonstrating the significance of government accounts came in 1977 when the Environmental Protection Agency (EPA) asked CSC to coordinate all its data processing. That same year CSC was awarded a contract by the Communications and Instrumentation Support Services at the Kennedy Space Center, thrusting CSC into the middle of the Space Shuttle program. Government contracts were important for two factors. First, the larger ones ran into the millions of dollars. Second, they were projects lasting several years or more, which meant a steady flow of money into CSC. Other contracts were for more specified projects and included writing medical applications, specific end user projects, and providing time-sharing—all of which drove up revenues to $452 million in 1980.

In addition to its regular offerings in programming and systems integration, CSC added others. For example, in 1978 it brought out a data base management system* called MANAGE. It also made it possible for end users to access databases through INFONET using the MRI System 3000 and software from TRW.† By the end of the 1970s CSC had a distributed network using specially built PDP–11/23 and PDP–11/44 miniprocessors and INFONET. Another version of MANAGE also operated on computers built by Digital Equipment Corporation (DEC).†

By the early 1980s CSC had 13,000 employees and annual sales of over $700 million. In 1984 it was the thirty-ninth largest company—as measured by revenues earned—in the data processing industry. Often characterized as "a grandaddy" among the systems houses, it had an image of rapid growth in the 1960s, moderate successes in comparison to the earlier period during the 1970s, and solid attainments in the 1980s. In 1983 revenues reached $718.9 million and declined to $709.6 in 1984, with earnings going from $15.9 million in 1983 to $27.6 million the following year. That was due in large part to CSC selling off a subsidiary called PAID Prescription, Inc. In that same period of time, most software houses generally experienced weakened sales. In large part, it appeared that CSC suffered from a decline in the number and value of contracts with the government. Yet by the mid–1980s it was still perhaps the largest software manufacturing company in the data processing industry. There were vendors, such as International Business Machines Corporation (IBM),† which sold a great deal more, but among those whose services were only software, CSC was one of the most important.

For further information, see: Computer Sciences Corporation, *CSC News* 15, no. 3 (April 1984): entire issue; Franklin M. Fisher et al., *IBM and the U.S. Data Processing Industry: An Economic History* (New York: Praeger Publishers, 1983).

COMPUTERLAND CORPORATION. This corporation was one of the largest of the early chains of retail stores specializing in computers. It became the channel through which hundreds of thousands of microcomputers were introduced into the U.S. economy during the 1980s, particularly for International Business Machines Corporation's (IBM's)† Personal Computer (PC) which became the de facto industry standard. The rise of such stores represented a subset of the data processing industry which, during the late 1970s and especially by the mid–1980s, had become a highly visible and obvious sector. ComputerLand grew rapidly from its establishment in 1976 to 545 stores in the United States and an additional 137 elsewhere by the summer of 1984.

In large part, ComputerLand was the creation of William Millard (1932-), and, like so many others who established businesses within the data processing industry, his beginnings in California did not signal the important future he would have. He was born in Denver, Colorado, in 1932, and, at the age of three, moved to Oakland, California, where he was raised. Later, he spent three semesters at the University of San Francisco where he decided that being a student was not his strength, particularly during a period when he could ill afford tuition. Earlier, between high school and college, he had been a ditch digger, worked in a copper mine in Nevada, and for a while worked at Southern Pacific Railroad in San Francisco. While working at Pacific Finance he collected bills and learned about business. In 1958 his company acquired a UNIVAC* whereupon he was offered the opportunity to work with it. Three years later, he became a data processing manager at Alameda County's data center. In his new capacity he developed the Police Information Network (PIN) which police departments in various parts of the United States in time adopted. After spending the period 1961 to 1965 at Alameda, he joined IBM as an Industry Specialist for State and Local Government, left a year later, and became the data processing manager for San Francisco County. In 1969 he began his first company, called System Dynamics, to produce systems software in competition with IBM; he failed and the company went out of existence in May 1972. He next formed IMS Associates for the purpose of bidding on a contract to design a fingerprinting online data retrieval system and won the contract jointly with TRW.† In the process of complying with another project involving a dealer for General Motors, he became familiar with the MITS microcomputer (1975) and by that exercise learned about micros in general. He had already acquired considerable knowledge about business in general and management through his various jobs and ventures.

Millard decided to sell micros himself, built at a facility he established, and called his first product the IMSAI 80. After some initial advertising, he quickly received orders for 3,500 machines, each of which was worth about $700. He chose to market the product primarily to commercial customers who represented a large, essentially untapped market. He also competed directly with Altair, which at the time was one of the giants in micros. Between 1975 and 1978 he sold over 13,000 machines, which made him a larger supplier than even Altair. Hence, he had now become one of the major forces in the American microcomputer market. Yet he experienced manufacturing problems, and there were quality control issues which, in 1979, quickly forced him to stop manufacturing such devices.

Millard now chose to operate franchised stores to sell microcomputers, using a formula evident at McDonald's hamburger chain. He also obtained his micros from various manufacturers. ComputerLand officially came into existence on September 21, 1976. Ed Faber, a friend of Millard's for some time, became president and the outward symbol of the company's leadership; Millard worked quietly out of the limelight. They opened their own and first store in Hayward, California, in November and the first franchised facility in February 1977 in Morristown, New Jersey. IMSAI Associates served as the holding company for ComputerLand. The two men subsequently expanded their franchises (each of which cost about $75,000 for a prospective owner) into hundreds of installations. When IBM announced its PC in the early 1980s, ComputerLand obtained a contract from the giant company to carry its PC products in these stores. A similar contract was negotiated with Apple Computer† and with dozens of vendors supplying peripheral equipment, software, and publications. The basic arrangement (as of the early 1980s) required that a store manager pay ComputerLand Corporation 8 percent of the gross sales in exchange for doing business under that name. These dealers also benefited from the deep discounts enjoyed by the corporate parent in acquiring inventories.

The formula proved to be a huge success. Quality of service, well-managed retail operations, and inventory on hand made ComputerLand a premier outlet for microcomputers at a time when such products were enormously popular. By 1982 the company grossed $25.5 million annually just in franchising fees, which led to profits of $10.8 million. In 1984 all these stores combined generated sales of $1.8 billion, of which an estimated $144 million went to ComputerLand Corporation. In 1985 sales continued up with an estimated $30 million profit, or a threefold increase over the previous three years. The company remained a privately held enterprise, leading industry watchers to estimate that Millard was worth between $500 million and over $1 billion.

The company's success was largely due to the rise of the microcomputer. From industrywide sales of several hundred thousand per year in the early 1980s to several million per year by the mid–1980s, it was, in hindsight, almost inevitable that retail dealers would do well. That fact was not so obvious at the time. Yet it turned out to be a particularly lucrative business for large dealers who franchised inasmuch as it called for no capital investments of any significance on the part of the parent firm while the enthusiasm for micros led many to take on the responsibility of operating a shop. As with so many others in the data processing business who were successful, progress came quickly for Millard. Barely half a decade elapsed between his financial destitution and his becoming a millionaire in 1980.

For further information, see: Paul Freiberger and Michael Swaine, *Fire in the Valley* (Berkeley, Calif.: Osborne/McGraw-Hill, 1984); Robert Levering et al., *The Computer Entrepreneurs: Who's Making It Big and How in America's Upstart Industry* (New York: New American Library, 1984).

CONTROL DATA CORPORATION (CDC). CDC, one of the oldest computer manufacturers in the data processing industry, was established in 1957 by ex-employees of Sperry Rand Corporation† in order to sell computer systems and do consulting work for the federal government. Over the next quarter century it evolved into an important manufacturer of peripheral equipment and intermediate-sized computer systems, sold products directly to end users and to various computer vendors, such as National Cash Register Company (NCR)* in 1960, which in turn were sold to end users. CDC became an important producer of International Business Machines Corporation (IBM)† plug-compatible hardware by the mid–1970s, a position it lost in the mid–1980s.

CDC was established in mid–1957 by William Norris and was capitalized at nearly $600,000. His first staff came from Sperry Rand; one of them was Seymour Cray who built a supercomputer named after him in the 1970s. CDC was headquartered in Minneapolis, where a staff of twelve designed and built CDC's first machine, the 1604 system, which they had announced in April 1958. They characterized it as the first solid-state computer. Taking advantage of a growing demand for computer-based expertise by the federal government, the company manufactured components for missile and aircraft systems and, on the civilian side, an air traffic control inquiry-keyboard-display terminal.

In its first year of operation, the company signed a half million dollars in contracts with the U.S. military establishment. Management used this money to complete the development of the highly publicized, advanced technology needed for the 1604. The first copies of the 1604 were shipped in January 1960 at a cost of just under $1 million each. Peripherals came from Ampex, Analex, IBM, and Ferranti Ltd.† In 1962 CDC brought out the 1604-A which supported COBOL*—then the industry's newest major programming language*—for use in commercial applications. The 1604 launched the company in its early life, but later it faced heavy competition from IBM and its 7090, 7044, and 7040. In December 1959 CDC had brought out a second system called the 160 and made first shipments in May 1960. With the 1604 and the 160 CDC had two products which were largely aimed at the scientific and engineering communities. It also made an arrangement with NCR to have its marketing organization sell 160s, and Ferranti did the same in Great Britain for the 1604.

In 1960 CDC began manufacturing peripheral equipment, and its initial product was a magnetic tape handler. In that year the company also decided to offer data services, selling time on its systems like other service bureaus. By 1965 it had seven such centers in operation which were used primarily by scientific and engineering customers and by CDC's own development staff.

Like other growing data processing companies of the period, CDC also sought to expand through acquisitions. It bought small companies that had specialized technologies or products that could enhance CDC's own products and position within the data processing industry. Cedar Engineering, acquired four months after CDC came into being, had the knowledge necessary to produce peripheral products. Then in 1960 CDC bought Control Corporation which could build computerized products for automatic control, particularly for use by electrical and gas utilities, and oil pipeline firms. Between sales of its products and acquisitions, CDC grew. From 12 employees in 1957 it went to 250 by mid–1958 and to over 1,350 in March 1961. Sales in 1958 were $600,000 and in 1961 reached $8 million. The firm proved profitable each of these years except during the first. Sales also kept rising: $63 million in 1963, $160 million in 1965, $245 million in 1967, and, in 1970, $1 billion.

New products helped fuel this growth. In July 1962 CDC announced its first large computer, called the 6600, which retailed for $7 million. The first one was delivered in September 1964. In May 1962 CDC announced the 3000 series with the initial model called the 3600. The Lawrence Livermore Laboratory took the first one in 1963 to use until it could get a CDC 6600. The 3600 was brought out to compete against IBM's 7044, 7040, and 7094, and, while it sold, it did not seriously reduce IBM's sales.

Norris kept acquiring other manufacturing facilities at the same time. In 1963 alone, he bought seven organizations. The most important of these were the computer part of the Bendix Corporation. Since the early 1950s Bendix had been a high-technology company that produced products and components for aviation, the automotive industry, ships, telecommunications, and so on, and had also entered the computer market. It actually built two machines, the G–15 and G–20 which were general-purpose computers, participated in the SAGE* project, and in 1963 had data processing revenues of some $13 million. But the firm was not committed to the massive investments and total concentration of efforts that it would take to be competitive and thus sold off its data processing operations. That same year CDC bought MEISCON which specialized in the use of computers to design highways and automate various industrial procedures. Norris purchased Beck's which made printing circuits and Electrofact which built and sold measuring, recording, and control equipment. Norris acquired expertise in terminals through the acquisition of Digigraphic from Itek. That same year the Control System Division of Daystrom swept into CDC, bringing with it the ability to manufacture electronic digital computers* aimed primarily at users in power, chemical, and petroleum companies. A small acquisition involved Bridge which made card punches and readers, as well as other miscellaneous peripheral equipment for use with computers.

CDC had committed itself fully to the data processing industry from its inception and aggressively set out to build an organization with the necessary expertise and facilities to develop, manufacture, and sell products. Norris believed that these products had to grow in size and breadth, and remain current technologically. The results of his strategy were reflected in the sales statistics cited above. Equally important was his idea that in order to grow and be successful in this industry management had to be willing to take risks in investments and inventions. The consequence was a successful company from its birth to the middle of the 1960s and beyond. Assets increased from $1,223,311 in 1958 to $71,338,765 by mid–1963.

In the early 1960s CDC, which had sold primarily to scientific and engineering communities, also realized that the distinctions between those groups and commercial users of data processing were blurring. Therefore, it needed to expand its offerings into the more traditional commercial market, home grounds for most computer giants such as Burroughs, NCR, Sperry, and IBM to mention a few. Between 1964 and 1968 CDC added commercial functions to its 6000 series of computers. These included a compiler for COBOL,* sort/merge software, and better file management packages. CDC broadened its 3000 series from its initial scientific focus, allowing it to be sold to commercial users as well. Such changes made it possible for customers to do both scientific and commercial processing in the same machine and thus gave CDC an early opportunity over other vendors to compete against IBM's processors which also could do both.

As the company grew in size, it felt the need to manufacture its own peripheral equipment; that requirement explained why some of the acquisitions listed before were made. Building and selling peripherals represented a very profitable venture. There was the opportunity both to make equipment to attach to CDC's processors and also to sell devices to other equipment manufacturers (which in the industry are called OEMs—original equipment manufacturers). In time, the capability allowed CDC to manufacture products that were plug-compatible with IBM's offerings, meaning that CDC's equipment could operate with widely used IBM computers.

Products poured out of CDC. In 1965 it bought Data Display, Inc., giving the firm the capability of producing more terminals. That same year it brought out the CDC 852 disk drive which was similar in design to the IBM 1311, and that was sold to General Electric (GE)† and Honeywell.† In 1966 the CDC 9433/34 disk drive came out; although it mimicked the IBM 2311, it was not fully compatible with it. The first shipments were made in 1967, clearly establishing CDC as a manufacturer of disk drives. By now CDC was selling OEM equipment to GE, Radio Corporation of America (RCA),† Honeywell, NCR, and International Computers Ltd. (ICL) in Great Britain. It even supplied products to the German computer firm Siemens. The 9433/34 was very successful; over 16,000 were ultimately sold. CDC also built the 6638 disk file for its OEM market. In 1967 new products were introduced for computer systems in general: a printer, tape transports, card read/punch gear, a magnetic drum storage unit, and various display terminals (CRTs). The following year a new disk file appeared that could hold 5 billion bits and a printer with a rated speed of 1200 lines per minute. In the late 1960s more tape and disk drives, printers, terminals, and card input/output came out. Thus, by the end of the decade CDC was producing all the peripherals one typically saw in a computer system.

In the 1960s CDC did not ignore its data centers. Between 1963 and 1969 these were expanded. In 1964 there were six which rapidly grew to seven. They used the CDC 3600 and 1604-A systems linked in a network called CYBERNET which used telephone lines. Scientific applications, such as operations research, ran with engineering usage, such as traffic surveying and planning. Other applications include administrative systems employed by hospitals and even scheduling of classes and logging grades for elementary schools. On June 30, 1968, CDC acquired a company called C-E-I-R that did programming and could perform other services as part of the company's overall attempt to expand the breadth of its service bureau operations. At the end of 1969, CDC had forty data centers throughout the world, used over 13,000 miles of telephone lines to work with these data centers, and installed a large collection of software that users could run in these machine shops. Yet because of enormous investments required to establish these facilities, CDC did not turn a profit in this part of the company until 1972.

In looking at CDC's enormous success and growth during the 1960s, it was obvious that, in addition to the announcement and shipment of many new products that were competitive in price and function, a large part of the story involved acquisitions. Between 1963 and 1969 CDC picked up forty-three different companies. They had cost CDC over $897 million. Franklin M. Fisher, a student of the data processing industry's economics, notes that thirty-eight of these firms made data processing products when they were bought. CDC worked to integrate these operations into a cohesive whole and did it more successfully than many other firms in the industry that also tried to grow by acquisition. The range of acquisitions proved significant, covering products for utilities, all major components or peripherals, optical character reading terminals (useful in retail marketing), service bureaus, programming, analog and digital computing, and scientific and commercial applications. It even bought companies outside the United States which in 1966, for example, included a portion of General Precision's Librascope Group and a components manufacturing firm in Hong Kong and a service bureau in Italy. In 1968 Pacific Technical Analysts, Inc., joined the flock at about the same time as C-E-I-R.

With all of these acquisitions, it was becoming obvious that tasks could be integrated into the company otherwise acquired through the services of vendors. In an attempt to drive costs down while taking advantage of new facilities, CDC periodically brought in-house functions that contributed to profitability or control by the company. For example, in 1966 CDC began manufacturing integrated circuits rather than continue buying them from such firms as Texas Instruments.* That kind of decision, when well executed, further encouraged the company to use its own resources. In that same year, it brought in-house the manufacture of card module assemblies, memories for computers, and other components used in building computers.

While these tasks took on new importance, other acquisitions were being negotiated. One of the most important of these came in August 1968 when CDC bought Commercial Credit. This firm financed, loaned, leased, and insured capital equipment. It was a financial institution that cost CDC $745,573,704 in stock—the biggest acquisition it made during the 1960s. CDC intended that Commercial Credit provide a vehicle for financing computer leases, and, in the decade to come, this arm of CDC played an important role in the overall performance of the firm. Yet this was not the company's first financial project. Since the early 1960s customers had demanded the capability of leasing equipment as opposed to just purchasing it. Therefore, CDC had to finance leased inventory and, in 1966, negotiated a contract with Leasco by which Leasco bought CDC's products and then leased them to CDC's customers. The arrangement did not prove satisfactory to CDC and was done for only one year. The experience probably encouraged CDC to make its acquisition of Commercial Credit which it could more fully control.

Although the 1970s did not represent as dramatic a period of growth as the 1960s, CDC continued to be a major player in the data processing industry. It began the decade with a broad and relatively current product line. It was shipping new computers, the CDC 6700 and 6200, in 1970. CDC continued to aggressively announce new products throughout the decade. Its only significant failure in this area was the Star 100 which in 1970 the company said would be a system capable of manipulating 100 million instructions per second. It built three copies of the computer before withdrawing the product in 1974 after missing delivery schedules and experiencing production and financial difficulties associated with it. CDC did make a major product announcement in March 1971, however, when it announced the CYBER 70 family of computers. As with most computer systems, this new one was less expensive than its predecessors and had more capacity and performance. In effect, it made CDC's 6200, 6400, 6600, and 7600 processors obsolete.

CDC's OEM business also enjoyed growth and strength. In 1972 CDC was the largest manufacturer of OEM products in the industry. That year, for example, it had about 150 customers made up of computer and peripheral manufacturers, and systems integrators. In 1970 CDC took the important step of deciding to manufacture peripherals that were plug-compatible with IBM's, insuring a head-on competitive battle between the two firms. Its initial plug-compatible manufactures (PCMs) were disk drives, such as the 23141 disk system aimed at displacing IBM's 2314. In 1971 Telex* and CDC began manufacturing IBM 3330-compatible disk drives which Telex sold to IBM's customers. With Fabri-Tek, CDC also produced "add-on" memory for IBM's processors. Even its lease contracts began matching or exceeding the attractiveness of IBM's and were designed to compete against the offerings of the larger vendor.

As the decade ground on, each time IBM brought out a new peripheral device of some significance, CDC soon after also announced one that was compatible. That practice caused industry observers to observe that CDC and other PCM manufacturers were simply "retro-engineering." That issue became very emotional because some manufacturers, such as IBM, suspected that CDC would buy a copy of a product, take it apart to see how it was built (in other words, backing out the engineering), and then with slight modification bring out their own version (hence the term *retro-engineering*). Because such a pattern of behavior by definition meant that CDC's equivalent products came out after IBM's, profits on these were not as high as they might have been otherwise for CDC. This was particularly the case with its 23141 subsystem and again in the 1980s with the 3380 look-alikes. Telex did not like CDC's 3330 PCM and canceled its contract while CDC ran into similar problems with Fabri-Tek. But with top management's renewed commitment to the PCM market in 1972–1973, CDC itself introduced more products aimed directly at IBM and sold by its own sales force.

The U.S. recession of 1970, which severely hurt many companies in the data processing industry, did not leave CDC undamaged. The company turned in a loss of $46 million in 1970 with sales down on computers, heavy writeoffs for projects that were not successful, but that made it possible for the company once again to be profitable beginning in 1972. Sales thereafter climbed with profits. In 1973 sales amounted to $948.2 million and approached $2.3 billion in 1979. This compounded growth rate of 15.9 percent exceeded that of the industry as a whole. Much of the success implied by these revenues came from peripheral equipment which, as a segment of CDC's business, expanded the fastest during this decade. Between 1975 and 1979, for example, sales of peripherals grew threefold. In 1975 they had only contributed $317 million in sales but in 1979 sales climbed to $909 million. As in the 1960s, its market for peripherals included users of CDC's computers, OEMs, and buyers of PCMs. Sales of PCMs to IBM's customers had grown, allowing CDC to claim in 1977 that over 1,300 data centers that used IBM's computers also installed PCM products. OEM sales went to 215 companies in 1973 but had grown to 720 by the end of 1977. Peripherals continued to cover the entire market: tape, disk, card, printers, and so on, including a broad collection of terminals.

In the 1970s CDC continued to experiment with joint ventures. For instance, in 1972 it joined with NCR to establish Computer Peripherals, Inc. (CPI) to build peripherals. In 1975 it linked with Honeywell to create Magnetic Peripherals Inc. (MPI) which made mass storage equipment. Some of the products made by such joint ventures, when combined with existing manufacturing facilities within CDC, allowed the company to introduce equipment aimed at IBM. It built memory for S/370* computers (from the Model 135 through the large 168) and, later, 303X processors, IBM-compatible disk drives to compete with 2314s, 3330s, 3350s, and 3380s. It went after IBM's 1403N1 printer (the most popular and widely used systems printer in the data processing industry during the 1960s and 1970s), but with only limited success. In 1973 part of its war with IBM ended with the conclusion of a lawsuit whereby CDC received the Service Bureau Corporation (SBC) from IBM.

During the 1970s CDC expanded its line of CYBER products. The 170 family grew, and in 1975 CDC brought out the CYBER 18, a small processor. In March 1977 two additions to the CYBER 170 family (called the CYBER 171 and 176) could perform distributed processing. In that same year CDC introduced the OMEGA 480-I and 480-II systems which were IBM PCM computers. Then in January 1978 CDC announced small CYBER processors. In April 1979 four more appeared called the CYBER 170 Series 700, which in effect replaced earlier members of the CYBER family. In June 1979, during the same month that IBM gave its customers their first shipping dates for on-order medium-sized 4300 computers, CDC announced more models of the OMEGA line which it claimed were faster than IBM's 4341 processor.

In 1980 revenues climbed to $3.8 billion, generating earnings of $260.3 million, and in 1981 they rose to $4.1 billion. Yet in 1982 they grew to only $4.3 billion. CDC made additional acquisitions in 1982: Computer Industries Corporation, Tabershaw Associates, Inc., and interests in Centronics Data Computer Corporation and Star Computer. Joining with fourteen other computer and semiconductor manufacturing firms, it established the Microelectronics and Computer Technology Corporation (MCC).

In 1984 CDC entered a period of considerable difficulty which extended at least through 1985. During 1984 earnings dropped by 80 percent as a result of changes in management and severe competition in computer markets. Declining volumes in sales of peripherals and service bureau work also hurt CDC. Earnings amounted to $31.6 million on sales of $5.03 billion, which still made it the fourth largest data processing company in the United States. It still shipped disk drives in large numbers, even though its market share was only about 29 percent, down from 50 percent in 1979. In 1984 the company decided to phase-out its PCM business which was no longer profitable. It announced its intent to sell off its Commercial Credit Company which also was not as profitable as it had been in earlier years.

For further information, see: Franklin M. Fisher et al., *Folded, Spindled, and Mutilated: Economic Analysis and U.S. v. IBM* (Cambridge, Mass.: MIT Press, 1983) and *IBM and the U.S. Data Processing Industry: An Economic History* (New York: Praeger Publishers, 1983); Stephen T. McClellan, *The Coming Computer Industry Shake Out: Winners, Losers, and Survivors* (New York: John Wiley & Sons, 1984); Carol Pine and Susan Mundale, *Self-Made: The Stories of Twelve Minnesota Entrepreneurs* (Minneapolis: Dorn Books, 1982); Donna Raimondi, "From Code Busters to Mainframes. The History of CDC," *Computerworld*, July 15, 1985, pp. 93, 98–99.

COOPERATING USERS OF BURROUGHS EQUIPMENT. See CUBE

CRC. See COMPUTER RESEARCH CORPORATION

CSC. See COMPUTER SCIENCES CORPORATION

CUBE (Cooperating Users of Burroughs Equipment). This organization was made up of companies that used computers manufactured by the Burroughs Corporation.† CUBE came into being in 1962 when CUE, a user's group of Burroughs B222 computers, and DUO, a user's group of the B205 and products made by Datatron, merged. In 1972 the organization was incorporated. Its primary mission was to serve users of B1000 and B7000 computers, representing their interests to the manufacturer while facilitating the sharing of experiences and information concerning these systems among members. CUBE, like many other user organizations in the data processing industry, served the needs of data centers with commonly installed equipment and software. At the start of 1980 CUBE had a membership in excess of 1,100 data centers. It held two annual conferences and maintained committees year-round to facilitate communications between members and with Burroughs.

For further information, see: T. S. Grier, "CUBE," in Anthony Ralston and Edwin D. Reilly, Jr., eds., *Encyclopedia of Computer Science and Engineering* (New York: Van Nostrand Reinhold, 1983): 429.

D

DATA GENERAL CORPORATION. Data General, established in April 1968, became one of the most important producers of a large variety of minicomputers during the 1970s. In an industry which in the late 1960s and early 1970s saw the birth of a new data processing manufacturing firm every three days on average, Data General was one of the few both to survive and to succeed spectacularly. It was created by a group of ex-employees from Digital Equipment Corporation (DEC)† who believed they could produce a low-cost, efficient minicomputer on their own and make money doing it. Their first product, called the Nova, was announced in late 1968 with first shipments made in February 1969. The product sold well, even against its primary rival, products from DEC. Originally capitalized at less than $1 million, Data General concluded 1970 with revenues in the United States of $6.8 million. Revenues grew along with almost instant profitability. During the 1970s Data General was usually the second most profitable computer manufacturer after International Business Machines Corporation (IBM)† which was consistently number one. By 1972 Data General's U.S. revenues grew to $25.8 million, and it closed the books in 1979 with total sales revenues of $507 million worldwide. By the end of 1983, it was the sixteenth largest U.S. data processing company, with sales of $867.1 million.

For its first sixteen years minicomputers represented its primary product lines although by the early 1980s, Data General was also selling peripherals, some software, and services. The Nova became the centerpiece of its earliest product set. A 16-bit processor, the Nova served as a small, general-purpose computer with up to 4,096 words of memory. It had adapters to allow attachment to dozens of types of peripheral devices, making it one of the most extremely versatile and, therefore, popular minicomputers. The base price of the original model was $7,950, one of the least expensive on the market. Over 700 copies of the original model were shipped by the end of its first year of production. During the 1970s additional machines that were larger and had more capabilities were introduced.

They could be linked together as well and had interfaces for linkage to the IBM S/360.* The first devices used core memory, but with the Supernova SC machines, monolithic technology became the norm.

In 1974 the Eclipse series was introduced and became another mainstay of the company for the rest of the decade. The entire product line found homes in factories where they were used for process control, time-sharing in networks, normal batch and online commercial applications, for database management, as front-end processors in large system configurations, as part of distributed processing networks, and in engineering centers. As demand for power increased and the cost of computing decreased, model introductions reflected market demands. In January 1978 the Eclipse M/600 was announced, priced at over a half million dollars per copy. This machine had up to 2 million bytes of memory and could be configured to house 6 billion bytes of disk storage, making the machine a large midrange processor by the standards of the day. Batch and time-sharing became possible with the advanced operating system (AOS) and one could write programs in PL/1,* the widely employed scientific language FORTRAN,* and the most broadly used commercial application language, COBOL.* XODIAC was also introduced in an attempt to manage networking better. In effect, users could link multiple systems together now or tap into other teleprocessing networks using Data General computers. The Commercial System (CS), introduced late in the decade, filled a gap at the small end of the product line, primarily for stand-alone business applications.

While business grew during the 1970s, there was severe competition. The introduction of new devices by International Business Machines (IBM),† for example, in the second half of the 1970s forced Data General to slash prices on its Eclipse series by about 50 percent. This was particularly the case in the face of IBM's 4300 computer announcement in February 1979. Sales grew slower in the early 1980s. The years 1981 and 1982 were weak sales years. Data General staged a comeback in 1983, with revenues up by 7.9 percent over the previous year and net income up 125 percent to $29 million.

The company was formed under the leadership of Edson de Castro who still ran the firm in 1984. Management, marketing an aging product line by 1981–1982, introduced new machines throughout 1983 which boosted sales. Its 32-bit Eclipse products were expanded in size and number of models, eating into DEC's sales (its most important competitor). The MV/10000 became the large-end machine and the MV/4000 the smallest, straddling DEC's VAX product line end to end. Two midrange models were announced called the MV/8000s. The company also introduced a microcomputer called the Desktop Generation with several models ranging in size from the Model 10 to the Model 30. The 10 was comparable to an IBM PC, while the 30 was a full-fledged minicomputer. Office products software also sold well, and large contracts were signed in 1983 with the U.S. Forest Service and E. F. Hutton.

For further information, see: Franklin M. Fisher et al., *IBM and the U.S. Data Processing Industry: An Economic History* (New York: Praeger Publishers, 1983); Tracy Kidder, *The Soul of a New Machine* (Boston: Little, Brown & Co., 1981).

DATA PROCESSING MANAGEMENT ASSOCIATION (DPMA). The DPMA is the largest data processing management organization in the world. Its purpose has always been to advance the interests and professional concerns of all types of data processing management through programs in education, research on self-improvement, and regional and local meetings. It strives to achieve the highest standards of professionalism possible within the data processing industry by actively promoting a formally stated code of ethics for all its members. DPMA also serves as a social gathering for people with common interests, reflecting a pattern of associational behavior that is characteristic of U.S. and Canadian societies.

DPMA was founded in 1949 as the National Machine Accountants Association in Chicago. By the end of 1950 it had over 100 members with chapters in New York, St. Louis, Detroit, Louisville, and Kansas City. On December 26, 1951, the organization received a charter of incorporation from the State of Illinois. That same year more chapters opened up throughout the Midwest, bringing the total number to twenty-seven.

The DPMA's first annual convention was held in Minneapolis in 1952 where it elected Robert L. Jenal, a systems manager at Toni Company in Chicago, as its first president. During these early years the official organ of DPMA, which started publication in 1950, was known as *The Hopper* and was one of the first nontechnical publications within the data processing industry. In 1954 DPMA changed its publication's title to the more serious *Journal of Machine Accounting Systems and Management*. The following year, DPMA established a chapter in Montreal—the first outside of the United States. By 1957 this worldwide organization claimed a membership of 12,000. Then in 1962, with its mission expanding as a reflection of the rapid growth then taking place within the industry in general, it changed its name to DPMA, having decided to expand its membership beyond those users of data processing associated with accounting and finance.

In 1962 the DPMA held its first examinations for the Certification in Data Processing (CDP)—an educational process that started two years earlier. It expanded educational offerings with the Registered Business Programmer examinations in 1970. Since then, exams have been given each year to members wishing such certification throughout the United States and Canada at various colleges and universities. Then in 1974 a separate organization was established for this purpose called the Institute for the Certification of Computer Professionals (ICCP).

In 1963 the organization's ponderously titled magazine was renamed the *Journal of Data Management*, and in 1966 a new publication appeared called

Automatic Data Processing Principles and Procedures. That year DPMA moved to its new headquarters in Park Ridge, Illinois. In 1968 it worked with the Boy Scouts of America to establish a merit badge program in computing and launched other activities in colleges and universities. DPMA ended a decade of phenomenal growth in 1970 by renaming its journal *Data Management*. Throughout the 1970s it carried out an aggressive publishing effort to define the principles of data processing management while expanding its offerings in the field of education in general with seminars and more frequent local programs. In 1974 DPMA became a member society of the American Federation of Information Processing Societies (AFIPS).† By 1980 DPMA's educational programs were widespread and included books, videos, movies, and lecture series, all of which were available to its members and could be used by others associated with AFIPS.

In the early 1980s DPMA became increasingly concerned with college and university programs leading to degrees in data processing. In 1981 it published a model curriculum for colleges to use as an example when constructing their own programs. Some colleges and universities have used this model as a benchmark against which to measure their own programs.

In the mid–1980s DPMA was a healthy organization with over 270 chapters in 1982. Each chapter had 25 to 500 members. In 1983 DPMA claimed a membership in excess of 45,000. At the top was an international organization with elected officers, followed by regional organizations around the world, each with fourteen to thirty-four chapters at the local level. New chapters kept opening up throughout the United States each year, insuring that the DPMA would continue to be one of the most important associations within the data processing industry throughout yet another decade.

For further information, see: I. L. Auerbach, "Data Processing Management Association (DPMA)," in Anthony Ralston and Chester L. Meek, eds., *Encyclopedia of Computer Science* (New York: Petrocelli/Charter, 1976): 427–429; Data Processing Management Association, *This Is DPMA* (Park Ridge, Ill.: DPMA, 1983).

DATAPOINT CORPORATION. This company was originally established as the Computer Terminal Corporation in San Antonio, Texas, on July 26, 1968. It came into being as a manufacturer of terminals and sold its products under the tradename of Datapoint. In December 1972 the name of the firm was changed to Datapoint Corporation. Its initial product was announced in 1969 and consisted of a solid-state terminal to replace old teletypewriter devices attached to computer systems. During the 1970s Datapoint broadened its product line to include processors, intelligent terminals, telecommunications gear, and software. The majority of these were intended for use in a distributed data processing environment.

During 1977 the company brought out the 6600 computer, a processor that could support up to twenty-four terminals and became part of many distributed processing networks. The next year Datapoint announced a small front-end processor that could be used with large International Business Machines (IBM)† mainframes. The company was always interested in telecommunications and as early as 1976 had merged data processing and telecommunications in a series of products which it called Infoswitch. That portion of the company involved in the manufacture and sale of that product line was sold off to Teknekron Industries, Inc., in 1983. Datapoint was the first company in the United States to have a local area network (LAN) product which it named the Attached Resource Computer (ARC) system. It built on this foundation over the next several years. One of the more interesting enhancements came in 1981 with an integrated circuit chip* that connected processors to a local area network. The following year it made these work with non-Datapoint computers. The LAN products worked well and during 1983 the company shipped its five thousandth network.

The company grew rapidly. From small sales in the early 1970s (for example, $34 million in 1974), the company closed 1979 with revenues of $242 million. The significant increase in revenues was due in large part to Datapoint's products in the area of distributed processing. These products were introduced in those years that saw an enormous expansion in the use of this kind of computing in general by large organizations. Revenues were substantial, with sales of $318.8 million in 1980, $508.5 million in 1982, and $600.1 in 1984. Net earnings grew in 1980 and 1981, dropped in 1982, but then increased fourfold in 1983, and again by threefold in 1984.

Building on its expertise as a manufacturer of networking products, in 1984 the company allowed other vendors to see its ARC LAN in detail, thereby making it possible to attach to it products that were doing very well in the general marketplace, including IBM's Personal Computers (PCs). At the end of the year, the company had operations in over forty countries besides the United States, its home base.

For further information, see: Pamela Archbold and John Verity, "The Datamation 100," *Datamation*, June 1, 1985, passim.

DATAPRODUCTS CORPORATION. This company, a subset of the manufacturing wing of the data processing industry, constructs peripheral equipment for computer systems. Its main customers are suppliers of computer products. Data Products Corporation, as it was originally known, was not only one of the first such firms within the industry, but also became best known as the world's largest manufacturer of printers for computer systems.

The company was incorporated on March 1, 1962, by a team that had worked together at Telemeter Magnetics and at Ampex, with Erwin Tomash as president. The initial core of the firm came from the Data Systems Division of Telex Corporation.† Programming services were settled into a subsidiary called Informatics, Inc.,† which was put under the management of Walter Bauer. By the end of March 1963 it had introduced a random-access data storage device called the Discfile and a printer with a rated speed of 300 lines per minute. Both products had been under development at Telex. Revenues in 1963 amounted to $3.176 million, and the firm had 241 employees. That year the company also lost $829,180. Additional printers (such as the 4100) and more file products emerged, while the key customers were original equipment manufacturers (OEMs). In 1966, when the firm began manufacturing computer memory, revenues reached $11.477 million, and the company had 578 employees. To broaden its product offerings, Data Products acquired Uptime Corporation, a card punch equipment manufacturer, in 1967. As for the rest of the data processing industry, 1968 was a boom year for Data Products: its sales went up 59 percent and earnings 800 percent. That year the company bought Fairchild Memory Products, and additional plant capacity outside the United States helped boost the number of employees to 1,300.

Additional acquisitions in the area of telecommunications helped bring 1969's revenues to $36.397 million and the number of employees to 1,929. With the recession of 1970–1971, Data Products's revenues increased 25 percent but its income went down 85 percent. The company began to make encoding equipment for the manufacture of credit cards and continued to add new printers to its product line. By now the firm was also leasing equipment, especially printers, to end users directly. In 1971, with revenues not growing, the leasing end of the business declined and telecommunications experienced an outright loss. In the early 1970s printers began to dominate Data Products' offerings, with core memories making the company the third largest manufacturer of such components in the industry. By the end of 1973 it was second only to International Business Machines (IBM)† in the manufacture of printers.

In 1973 Dataproducts merged the two words of its name into one, and its revenues approached $76 million. The company also shipped over 3,000 printers, a product that would dominate the company's earnings for the rest of the decade. In 1977 it manufactured 10,503 printers. Revenues in 1979 reached $163.569 million, and in 1980 IBM began purchasing printer-related products from the firm. Total sales for the company hit $180,319 million, making it one of the most important manufacturing companies dedicated to the production of all sizes of printers. The following year, the company's sales were more than 50 percent over the previous year's, increasing to $270 million, while earnings rose by 134 percent. Such results were typical of companies offering a new technology. In the case of Dataproducts, its entire product line in printers was broad, relatively new, and thus cost-effective with appropriately competitive performance, and the company had begun selling in the office products market. In a 1984 survey of all data processing companies, *Datamation*, a leading journal in the industry, ranked the company forty-eighth from the top in annual sales. Revenues that year exceeded $484.5 million. By then the largest manufacturer of printers in the industry, it turned in yet another record year of growth with revenues up 41 percent over 1983. IBM continued to buy products from the firm, and major contracts were signed with the Burroughs Corporation† and Wang. As in previous years, the company continued to expand its manufacturing capability and capacity, introduced new products, including more supplies and components, while selling on a worldwide basis.

Its specialty was clearly printers and had been so for years, despite other acquisitions and later divestments of unprofitable ventures. It began 1985 as a strong firm with approximately 5,800 employees and a solid position in the industry. The history of this firm offers a window into the evolution of printing technologies, but it also illustrates the belief of many businesses that success comes to those who do one thing well.

For further information, see: Dataproducts Corporation, *The First 20 Years of Dataproducts* (Woodland Hills, Calif.: Dataproducts Corporation, 1982); Rick Forman, "Tales in Peripheral Enterprise: The Rise of Dataproducts Corporation in the Computer Industry, 1962–1972" (M.A. thesis, University of California, Santa Barbara, 1980).

DEC. See DIGITAL EQUIPMENT CORPORATION

DECUS (Digital Equipment Computer Users Society). DECUS was established in 1961 to facilitate sharing of information about Digital Equipment Corporation (DEC)† equipment among users and to convey the requirements and experiences of that company's customers back to DEC. Almost every major community of computer users has its own organizations, for example, GUIDE† and SHARE† for users of International Business Machines' (IBM's)† computers. Thus, DECUS reflects a well-established pattern within the industry. In 1980 it had about 35,000 members, making it one of the largest user groups in the industry. The size reflects the fact that DEC sold minicomputers to thousands of end users, such as engineers and scientists, rather than just to data centers. DECUS was unique in that, unlike other user organizations, it developed a library of software packages that ran on DEC's equipment which members could acquire. By the early 1980s that collection of programs exceeded 1,700, covering primarily a wide range of scientific and engineering applications. DECUS holds two annual conventions at which DEC introduces new products and where members deliver papers.

For further information, see: C. Lickteig, "DECUS," in Anthony Ralston and Edwin D. Reilly, Jr., eds., *Encyclopedia of Computer Science and Engineering* (New York: Van Nostrand Reinhold, 1983): 520.

DIGITAL EQUIPMENT COMPUTER USERS SOCIETY. See DECUS

DIGITAL EQUIPMENT CORPORATION (DEC). DEC developed some of the most widely used minicomputers of the 1960s and 1970s, known as the PDP series. The widespread use of minis was the direct result of DEC's decision to build small, low-cost computers that were easy to use. That decision introduced the concept of having one computer for one application and furthered the process of "distributing" applications to user communities with small processors. Like many other companies formed in the late 1950s that survived over the next quarter century, DEC specialized in one major area, creating its own market and, particularly in its early years, avoiding major competition from International Business Machines Corporation (IBM),† Burroughs,† Honeywell and other large vendors. Its products were technologically advanced and continually broadened the computer marketplace. Although by the 1980s its product line was broad and it now competed heavily against large computer vendors, its history is one of success and growth. Today DEC is one of the major computer manufacturers of minis and related equipment.

DEC was formed in 1957, the same year as Control Data Corporation (CDC),† by Kenneth H. Olsen** and a friend, Harlan Anderson. The dominant force behind the company, however, was Olsen. He wanted to build inexpensive, fast computers that were easy to use and easy to manufacture. He started the company with $70,000 provided by a ventures capital group called American Research and Development. Olsen dominated all aspects of the new company's development and soon acquired a reputation as one of the industry's best executives. He majored in electrical engineering at MIT and, while a student there in 1950, worked at the Digital Computer Laboratory at the university on the Whirlwind Computer,* a job that gave him considerable insight into computer design. After the Whirlwind project was completed, the U.S. government awarded MIT management responsibility for the SAGE* defense system while IBM won the contract for the construction of its computer equipment. As part of the SAGE project, Olsen went to IBM's manufacturing facility in Poughkeepsie for two and a half years. There he learned more about how computers are designed and moved from the laboratory through manufacturing. From that experience he concluded that he could make computers more efficiently while he gained a good appreciation for production issues. He also acquired a good understanding of IBM's policies and practices in personnel management which he later carried over to DEC.

At the end of his tour of duty in Poughkeepsie, Olsen was thirty-one years old. He had experience as a designer of computers, knew how to manufacture them, was learning more about budgets, accounting, and finance, had been a manager, and was ambitious. It was also 1957, the year he formed DEC. After obtaining the capital he needed, he set up his first headquarters in an old woollen mill in Maynard, Massachusetts. His staff grew very slowly, and he did not hire the company's first salesman until nine months after the company was started. DEC's salesmen were engineers, and, unlike their peers throughout the industry, they were not on commission.

DEC's original products were not computers. It began by making circuit modules which other companies used in test equipment or other computers. Later, these modules became important building blocks in DEC's own computers. Meanwhile, Olsen was developing a small organization that was designing its first small, yet fast, computer.

Olsen introduced his first computer to the industry in 1960: the Programmed

Data Processor, or simply the PDP–1. This computer, like many that followed it, was intended for use by the scientific and engineering community. It sold for $120,000, nearly $800,000 less than rival devices. Because his customers were knowledgeable about data processing, Olsen did not have to offer considerable support services to these people as IBM and others had to with commercial customers who had little knowledge of data processing. This important factor, as with CDC's customers, made it possible to keep costs and therefore the price of computers down. Because the sales force was made up of engineers selling to fellow engineers, from the very beginning Olsen could appreciate what his customer set wanted. The result was instant success for the PDP–1 and its follow-on products.

The PDP series became the most popular computer for engineers during the 1960s and 1970s. Throughout the 1960s, Olsen produced other PDPs. They became faster and less expensive. One of the most important was the PDP–8, introduced in 1965. It was his first computer made with integrated circuits, and it cost only $18,000. It became one of the most widely sold members of the PDP family. Sales of these and other models caused DEC revenues to grow ninefold between 1965 and 1970. Net earnings increased by a factor of twenty in the same period.

Customers justified the purchase of PDPs on the basis of (1) their low cost relative to other available machines and (2) the applications run on them. These small computers were employed for data gathering, experimental work in laboratories, and later, in manufacturing plants to automate functions and gather shop floor data. Customers embedded them in process control systems and equipment or made them critical elements in a distributed processing network. Distributed processing became an important feature of the industry by the early 1970s. As the cost of computer power kept dropping, along with the size of these machines, organizations "distributed" their processing capability in various departments and locations without having all data processing work done on a centralized large computer. PDPs and other minis were also sold in the 1960s and 1970s for single applications as opposed to the more traditional approach of running multiple applications concurrently on a large mainframe. PDPs appeared in a variety of places where no computers had been before. They were in universities and companies, in military installations and on ships, in coal mines and laboratories, on plant floors and in hospitals.

One can gain an appreciation of how quickly the industrialized world seized on minis by examining DEC's sales record. DEC was the industry's most successful manufacturer of minicomputers. By 1966 it had 1,100 employees and 800 customer installations in six countries serviced by twenty-four sales offices. In 1970 DEC had 5,800 employees to support its sales and manufacturing and had customers in eleven countries. In order to raise capital for expansion to satisfy growing demand for its products, DEC made its first public stock offering in 1966 which raised $4.8 million; others followed in the late 1960s.

The company had developed a strategy that called for the continuous announcement of new products that technologically were at the forefront of technological innovations and, often, of price performance. All products had to be superior to those of their predecessors and their competition. These machines were characterized by continued growth in capacity and function, with high performance hardware from model to model as time past. DEC plowed back into research and development each year between 8 and 11 percent of its revenues to insure that new machines were continuously being introduced.

As a consequence, the record of product introductions was impressive. The first PDP came out in 1960. The PDP–4 appeared in 1962 and the PDP–5 the following year. The PDP–6, 7, and 8 came out in 1964, the PDP–8S, LINC 8, PDP–9 in 1966, and the PDP–8I and PDP–8L in 1968. In 1969 DEC announced the PDP–12, 14, and 15. The company ended the decade with the PDP–11.

The PDP–8 was a particularly important machine for DEC in that its popularity boosted sales revenues while becoming a technological landmark in itself. It had four times the speed of its predecessor, the PDP–5, and sold for one-third less. By the middle of 1966, over 400 of these machines had been installed. Other members of the PDP–8 family were introduced to broaden this product line into the mid–1970s. One of the last members of this successful family of computers was the PDP–8A introduced in 1975. Each new model was less expensive to purchase than older ones, and they were always faster and more reliable. These PDPs evolved from being simply scientific processors to general-purpose computers with considerable function and performance that could be used increasingly for business applications. Between 30 and 50 percent of all PDP–8s were sold to companies that incorporated these computers into their own systems or machines. The other half went to traditional users of computers such as companies, laboratories, and universities. By the fall of 1975 DEC had sold between 30,000 and 40,000 PDP–8s.

Also contributing to its revenues were sales of peripherals by DEC. DEC also developed some software, yet neither constituted a significant portion of the business. DEC introduced some disk drives, paper tape units, terminals, and communications controllers. It developed a programming language for business applications called DIPOL for the PDP–8 and, later, COBOL* for the PDP–10. It also wrote some numerical control software during the 1960s.

During the same decade, in addition to its strategy of offering technologically advanced products which were easy to use, DEC employed other tactics. It maintained a small service organization and offered few software packages when compared to such giants as IBM, Burroughs, or Radio Corporation of America (RCA).† The company continued to rely on its highly sophisticated customer set of engineers and scientists for its business as much as possible in order to avoid having to maintain costly service organizations and software development groups. Furthermore, it frequently sold machines to support one application, while companies such as IBM usually opted to propose large mainframes to handle multiple applications. Thus, decisions to acquire DEC processors were frequently made outside the data processing department by end users. Such decisions were also easier to make because they cost less.

DEC's business volume illustrates the success of these strategies. Assets rose from $5.7 million in 1964 to $114.8 million in 1970, and profits after taxes grew from $889,000 to $14.4 million in the same period. In 1961 revenues were $4.3 million, and in 1970, $142.6 million. That record represents a compound growth rate of 44 percent per year. In 1970 DEC had become the third largest manufacturer of computers after IBM and Sperry.†

Growth was spectacular in the 1960s, and it continued to be so during the 1970s. Between 1972 and 1979, revenues jumped from $188 million to $1.8 billion—a 38 percent compounded growth rate which was phenomenal for a company that size.

These sales were continuously supported with new products in the 1970s as they had been in the 1960s. The PDP–11D23 came out in 1972 as a front-end communications processor. It was marketed against IBM's popular 3705 telecommunications controller. The Data System 500 family of computers matured throughout the 1970s. In 1973 DEC announced virtual memory for its computers. This important function allowed disk files to be swapped back and forth in a computer's memory quickly, thereby expanding a processor's capability to handle larger quantities of data within one or more applications concurrently. Members of the PDP–11 family continued to sell well throughout the 1970s, along with those of the DEC system 10. Many of the computers of either family were marketed against IBM's S/370s,* from the small 115 through the 148.

DEC introduced a new product line in 1977 called the VAX–11/780, which was aimed at IBM's large 3031 and 3032 customers as well as engineering customers with large processing requirements. These were compatible with the earlier PDP–11 processors, thereby offering DEC's traditional customer set growth capability. DEC also offered compilers for FORTRAN,* COBOL,* and BASIC* on the VAX series, making these even more attractive for use within a commercial customer set than ever before. Additional members of this family of computers appeared throughout the late 1970s and into the early 1980s.

After IBM announced the 8100 distributed processor in 1978 and the 4300 computer family in 1979, DEC reacted like most of the industry by reducing prices for its PDP–11/70, PDP–11/34A, VAX–11/780, and DEC system 10. Such price reductions by the entire industry marked again a new level of price performance in which machinery became less expensive for the services they performed, causing more computers to be sold throughout the industry. Such sales, while they initially hurt DEC's profits per unit, insured that continued sales growth for all competitive computer manufacturers would continue.

The company had grown and prospered over a quarter century because it had a specialized product—the minicomputer—which it sold primarily to the engineering and scientific community. Although DEC competed against other large computer manufacturers such as IBM beginning in the mid–1960s, its market remained specialized. DEC continuously introduced technologically advanced computers that were sold much the same way as CDC did theirs: by small marketing groups who had engineering and scientific backgrounds. The PDP product line was the most popular family of computers with engineers in the 1960s and 1970s, and thus the product sold well. It made significant inroads into that community because most computer manufacturers were building processors for commercial customers. Not until the early 1980s did most large vendors focus renewed attention on the computing requirements of the engineering community—ironically one of the first groups to use computers when the data processing industry was born.

DEC was always a well-managed company and generously funded its own research and development. It was one of the darlings of Wall Street and met with success whenever it chose to raise capital. It entered the 1980s as a major vendor in the computer market.

For further information, see: "A Q and A with DEC's President Ken Olsen," *Computerworld*, June 7, 1982, pp. 10–11; Franklin M. Fisher et al., *IBM and the U.S. Data Processing Industry: An Economic History* (New York: Praeger Publishers, 1983); Katharine D. Fishman, *The Computer Establishment* (New York: Harper & Row, 1981).

E

ENGINEERING RESEARCH ASSOCIATES (ERA). ERA was one of the first commercial manufacturers of computers formed in the United States. Established in 1946 by former naval officers involved in communications and radar research during World War II, the company was set up to continue providing the government with computational equipment. William Norris was the first head of the firm. ERA's first project was to build a stored-program computer for the Navy called ATLAS I. A commercial version of that machine was named the 1101 and was marketed in 1948. The first 1101 was delivered in December 1950, and the first UNIVAC I* in March of 1951.

This vacuum tube-based technology was next modified to build the 1102 computer which had more general-purpose features and which was sold for $575,000 in 1952. The 1102 used teletype punches, paper tape readers, digital plotters, and printers. ERA sold three 1101s and three follow-on products. Despite limited sales, as early as 1949 ERA began research and development of the 1103, and the first copy was introduced in 1953. This machine did far better than the earlier ones: nearly twenty were sold.

In 1952 Remington Rand† acquired ERA and its 500 employees as a means of quickly acquiring technical skills in the area of computer design. The acquisition was important because in 1952, according to ERA, the organization "had built and delivered more than 80 percent of the value of electronic computers in existence in the United States at that time." The relationship did not work out, however, and within five years, some of the key personnel at ERA left to form Control Data Corporation (CDC).† Among them were Norris and Henry Forrest, two of the more important managers and designers of the 1101 and 1102.

For further information, see: Franklin W. Fisher et al., *IBM and the U.S. Data Processing Industry: An Economic History* (New York: Praeger Publishers, 1983); E. Tomash and A. A. Cohen, "The Birth of an ERA: Engineering Research Associates, Inc.," *Annals of the History of Computing* 1, no. 2 (October 1979): 83–97.

ERA. See ENGINEERING RESEARCH ASSOCIATES

F

FAIRCHILD SEMICONDUCTOR CORPORATION. This company, along with Texas Instruments,† was the earliest manufacturer of chips,* the basic electronic building block of modern computers. Fairchild also spawned many high-tech companies that collectively occupied a section of California later known as Silicon Valley. Fairchild became the parent of many companies which, beginning in the 1960s, would provide innovations in hardware for computers and a wide array of electronic gear ranging from rockets and weapons to computerized microwave ovens and video games.

Fairchild Semiconductor was established by an investment made by Fairchild Camera and Instrument Corporation. In the 1950s, its president, John Carter, became interested in acquiring a company that could manufacture solid-state transistors. In 1957 William Shockley,** a scientist who had done considerable research on semiconductors and who was already known as the father of transistors, approached the investment firm of Hayden Stone to find a sponsor to establish a new company. Hayden Stone represented Fairchild Camera. Negotiations resulted in a company that would make semiconductors based on silicon transistors rather than the more fashionable germanium. Fairchild's interest was spurred by the demand of its largest customer, the U.S. military, for further miniaturization of electronics.

Robert Noyce,** another physicist, joined with Shockley to establish Fairchild Semiconductor at Mountain View, California, near Palo Alto in what would one day be called Silicon Valley and in large part because of Fairchild. It was founded in 1957, the year that Sputnik galvanized the United States into renewing its commitment to another generation of new military hardware requiring more compact, efficient electronics. The parent company provided the venture capital necessary to launch it, with the option to buy out the entire company. This was one of the first times that scientists who had developed computerized technology on the West Coast aspired to make money off a project as opposed to simply living off a salary. The only other earlier example was the case of the builders of the UNIVAC† computers in the 1950s. The result was a multibillion dollar economic boom over the next quarter century.

The idea of miniaturizing electronics was not new, nor was Fairchild the only company that was seeking to move into different electronics. Texas Instruments also took a leadership role in this area, and several other firms, such as International Business Machines Corporation (IBM),† and even General Electric (GE),† made heavy commitments to computers. In the early 1950s, transistors were slowly replacing vacuum tubes in electronic equipment, increasing the reliability of such hardware and decreasing their size as well as prices. The work done on transistors led scientists to the idea of the integrated circuit (ICs), or the microchip, in which electronic functions were put together without benefit of wires and in more compact mass than was possible with transistors. This research was funded primarily by the military.

Fairchild Semiconductor wanted as much independence as possible in the beginning in order to have the freedom to strike out in new directions. Years later Noyce commented that only about 4 percent of Fairchild's R&D budget ever came from the military during the early years. Noyce and his colleagues initially focused on the manufacture of silicon transistors but later diverted their attention to eliminating the inefficient hand wiring of transistors. In 1959 Noyce, now director of research and development, worked with Gordon Moore, a chemical physicist, to find a transistor that would have no wires or moving parts. Their goal was to put an entire electronic subsystem on a silicon chip without wires. Similar efforts were underway at Texas Instruments under the guidance of Jack Kilby.** Both firms succeeded in 1959, evolving the transistor from solid-state technology to what we today know as the chip. Both companies filed for patents for their integrated circuits and then spent the 1960s in litigation over who really developed this technology. In 1969 a court ruled in favor of Noyce, even though the scientific community had earlier credited both men for the development.

Concurrent with these legal squabbles came the real revolution in electronics—the manufacture of integrated circuits, which by the mid–1960s produced another round of miniaturization in electronics and an explosive growth in new electronic equipment, including computers. Within a few years, the integrated circuit swept away vacuum tubes and transistors, along with solder and endless miles of wiring in many sophisticated pieces of electronic equipment. To a large extent, Fairchild also led the way in the early 1960s through efficient manufacturing techniques. It gave royalty permissions to other firms to take advantage of research it had done on ICs.

Fairchild's first major contract using chips was the Minuteman project, one of the earliest government efforts to build intercontinental ballistic missile (ICBM). Its designers used integrated circuits for the missile's electronic subsystems. That contract established Fairchild Semiconductor as a viable business venture. The decision to use ICs revolutionized the entire space program, creating almost overnight the new growth industry of semiconductors. The revolution was already evident by the end of 1963, when 10 percent of all electronic circuits in the United States were ICs. The new venture was profitable, and so Fairchild Camera exercised its option and bought back Fairchild Semiconductor in 1963. By the mid–1960s, Semiconductor had annual sales of $150 million.

Noyce became a millionaire, and his experience encouraged others. As the leading laboratory on the West Coast specializing in semiconductors, Fairchild exposed many scientists and engineers to the new technology and drew considerable venture capital into Santa Clara Valley. The result was a series of defections from Fairchild, creating many new semiconductor companies in what became known as Silicon Valley in the 1970s.

The first of these new firms was Rheem Semiconductor which was acquired by Raytheon† Semiconductor in 1961. That year other members of Fairchild founded Signetics Corporation and Amelco (which later became known as Teledyne Semiconductor). Defections continued as General Microelectronics came into existence which made metal oxide semiconductors called MOS chips. Unlike Fairchild's, these chips ran slower but used less electricity. Fairchild produced what was known as bipolar circuits which the military favored because of their high speed and used them extensively in missile guidance systems. For similar reasons, bipolar technologies were also widely employed in the American space program.

Fairchild prospered in the early 1960s. In 1965 the company had sales offices in various parts of the United States and plants in California and in Maine. In Hong Kong, cheap labor did assembly work, bonding chips together into packages that were smaller than one-inch square. Competition grew, and by 1965 twenty-five firms were in the same business, each stealing technicians from each other and especially from Fairchild. That year the company reorganized its marketing force into four distinct areas—military, industrial, computer, and consumer—confident that chips could be sold to each.

Industrial accounts were a good potential market because more complex electrical machinery could be built with chips. The military, already convinced of the benefits of ICs, became a major source of revenue for Fairchild and has remained so to the present. By 1970 Fairchild and other firms had penetrated this arena, and many computer vendors were making their own chips either for the military or for their own products. IBM, for example, had its own manufacturing facility at East Fishkill, New York, in the early 1960s. IBM's System/360,* announced in 1964 and the single most important computer product ever introduced within the industry, was based on the new technology. In one blow it wiped out demand for vacuum tubes or transistors in computers. Later, the same company became the first major vendor to build all its computer memories out of chips.

Ironically, sales to computer companies were slow in the beginning, largely because many vendors, IBM, Bell Laboratories† of American Telephone and Telegraph (AT&T),† Burroughs,† Control Data,† and Sperry,† were regarded as the innovators in electronics, and they had not yet developed chips. Until they became involved, others within the computer manufacturing community held back. Some scientists and engineers in these electronic firms also feared that firms like Fairchild could put them out of work.

The consumer market for chips proved to be a disaster for all concerned. The market (radios, televisions, etc.) was the most unsophisticated in the use of new technologies, particularly for television. Television sets continued to be made with vacuum tube technology, and companies such as GE and Radio Corporation of America (RCA)† were reluctant to press for a change. At this time the government began to insist that televisions be able to handle both UHF and VHF channels in large numbers, which current technology could not readily support. Tubes, for example, too frequently failed when used with UHF frequencies. TV manufacturers moved slowly, however, and instead went to transistors while computer companies were already moving to chips. But it was a start, and Fairchild was selling its products. In 1965 alone it sold nearly $6 million in chips for use in consumer products.

By the time Neil Armstrong landed on the moon in 1969, Fairchild was a leader in research and development in the field of semiconductors and had a good marketing organization. Its strategy was to develop silicon-based chips of such universal applicability as circuits that markets would develop around it. While old-line electronics firms resisted these changes (during 1961 twenty-nine had been found guilty of price fixing), a younger breed of companies were fiercely competing for new markets. A new generation of technically educated and experienced executives was rebuilding the electronics industry. Many of these leaders continued to come from Fairchild. That organization alone spawned three new companies making chips in 1966, three more in 1967, another thirteen in 1968, and eight in 1969. These companies were led by risk takers, entrepreneurial types with experience in making chips at Fairchild. They were lured away by the prospect of wealth and the excitement of developing new products on their own. Unlike the managers of the 1950s who wore suits, these scientists were casual, free thinking, unconcerned about institutions, and competitive to the point of ruthlessness. Most were under forty years of age and frequently were under thirty when they became millionaires.

Fairchild's great era waned in the 1970s as it continued to lose talent. Fairchild's California location was one reason for its problem. The fine weather and flexible lifestyle led a number of companies to spring up near Fairchild. The availability of additional talent from California Technological Institute and Stanford University, not to mention the University of California campuses all along the state coast, also encouraged the synergism of the new electronics industry. Venture capitalists were eager to establish new firms and often went after Fairchild's employees for the talent. Alumni also recruited others. Periodically, rather dramatic defections occurred. For example, Charles Sporck, who had been instrumental in making the use of planar integrated circuits effective, left the company in 1967 to become president of National Semiconductor. He had been a general manager at Fairchild, and his new firm became one of the most important chip and memory manufacturers in the industry during the 1970s.

Though Sporck's defection was one of many, it brought the problems between Fairchild Semiconductor and the parent company, Fairchild Camera, to a head. The company in California had lost too many good engineers along with the leading-edge technologies that they were developing. Others were becoming rich and successful with ideas first developed at Fairchild. Another problem involved the way the company was run. The independence and free-wheeling atmosphere that had been so inviting to scientists at Fairchild Semiconductor in the late 1950s had also eroded. As the company grew in size, the parent firm increasingly imposed restraints and guidelines intended to squeeze profits out of Semiconductor to support less successful divisions on the East Coast. Frustrations culminated dramatically in June 1968 when Noyce left Fairchild. But by then, Fairchild had little technical talent left.

Fairchild Camera responded with an aggressive program to lure back engineers. Other locations were raided, including Motorola's talented staff in Phoenix. There, under the management of C. Lester Hogan, almost the entire senior staff within Motorola's semiconductor operations left for Fairchild. But then Fairchild suffered another critical loss: its leading marketing expert, Jerry Sanders, who had headed up the company's sales to the military, resigned. Meanwhile, Noyce, his original co-founder of Fairchild, Gordon Moore, and another colleague, Andrew Grove, established Intel Corporation.†

Fairchild's great era had ended. In subsequent years, it continued to manufacture electronic components and, later, microprocessors (Intel's contribution and strength). It supplied military projects and some other firms in the electronics industry. In 1979 Fairchild Semiconductor was purchased by Schlumberger Ltd., a French company, for nearly $350 million. Fairchild had been the bellwether company of Silicon Valley in the late 1950s with innovative technology. In 1979 it unwittingly again served the same role. Through its acquisition by a foreign company, it represented the first major takeover of a high-tech company in the valley by a non-U.S. firm. This pattern would be repeated many times in the 1980s when European and Japanese investors acquired American companies.

Fairchild sold for a relatively inexpensive price because it had not done well in the late 1970s. Misjudgment, such as its heavy investment in digital watches, had particularly hurt the company. The market for such watches collapsed even before Fairchild could move people into the new facility it erected to house its Consumer Products Division—a multimillion dollar error.

Fairchild's history represents an important chapter in the story of how electronics influenced data processing technology. It was at the center of the most important innovations in electronics, computers, and technology in the mid-twentieth century. Although only briefly a central part of that story, the continued reliance on chips thirty years after silicon wafers were developed is a testimonial to the profound contributions of the founders and early leaders of Fairchild Semiconductor Corporation.

For further information, see: Dirk Hanson, *The New Alchemists* (Boston: Little, Brown & Co., 1982); T. R. Reid, *The Chip: How Two Americans Invented the Microchip and Launched a Revolution* (New York: Simon & Schuster, 1984); Everett M. Rogers and Judith K. Larsen, *Silicon Valley Fever: Growth of High-Technology Culture* (New York: Basic Books, 1984).

FERRANTI, LTD. Ferranti was one of the first British companies to manufacture and sell computers. Among its machines were the Mark I (not to be confused with the Harvard Mark I), the ATLAS,* and the Manchester MU5. The company first became interested in computers when nearby Manchester University began building them in the late 1940s. This well-established electronics firm underwrote research at the university beginning in the fall of 1948. After the Mark I had been developed, Ferranti began building them, selling nine copies between 1951 and 1957. The first computer commercially available in Great Britain was a Ferranti Mark I installed in February 1951 at the university. The later machine had more capacity than the earlier device built at the university in 1949. Ferranti's production of computers peaked when it built and sold the PEGASUS* series in the late 1950s and early 1960s.

Ferranti was able to develop machines quickly by supporting academic research on computers. In 1958, for example, it agreed to participate in the development of the ATLAS, one of the largest computers built during the 1950s. The company's primary technical contribution to the project consisted of writing the Atlas Supervisor, one of the first operating systems ever developed; it was drafted by David Howarth. When the machine was completed in December 1962, it was one of the most powerful in Europe. As a "supercomputer" of its day, demand for the machine obviously was limited. Meanwhile, however, Ferranti had developed the SIRIUS, a computer based on the transistor. Its first SIRIUS was shipped in 1960, two years after its first transistor-based rivals began appearing, namely, the Elliott 803 and the EMI EMIDEC 1100.

The company continued its long tradition of developing new electronics, a pattern established by its founder Sebastian de Ferranti (1864–1930). As early as 1948, executives at the firm had appreciated the commercial potential of the stored program digital computer. With the use of government-funded research facilities at Manchester University, Ferranti quickly became the largest manufacturer of British computers in the 1950s. In 1951—the first year it sold computers—it had no competitors; it only had potential customers who knew little or nothing about computers. Worldwide, there were fewer than two dozen firms attempting to sell computers.

Ferranti produced many machines in the 1950s. First came the Mark I, delivered initially in 1951; two were sold. In 1953 the Mark I Star appeared, and the company sold seven. In 1956 the PEGASUS 1 represented a major jump forward in quality. Twenty-three were sold in Great Britain, and an additional three were exported. The PEGASUS 2 (first shipped in 1959) resulted in twelve sales, whereas the earlier MERCURY system of 1957 accounted for nineteen other shipments. In 1959 the company also brought out the PERSEUS but sold only two. The PEGASUS series proved to be the most important because much of the technology and packaging developed for that machine influenced future processors from Ferranti, including the ARGUS, ORION, and FP6000. The PEGASUS was also the first computer used by any British firm in a service bureau. In March of that year Ferranti used the PEGASUS to establish its first service bureau, located in London. It was the most successful and reliable of Ferranti's early machines.

The PEGASUS satisfied those who needed very large machines. To take care of smaller users, the company developed the ARGUS, a series of machines that became available in 1963. For the military community in the 1960s, Ferranti built the POSEIDON, HERMES, and the F1600. The F1600 machine was eventually modified many times to satisfy customer needs, allowing the company ultimately to sell several hundred of them around the world.

The ORION was yet another machine under development at Ferranti around 1960. The engineers working on it experienced considerable technical difficulties with this large device, particularly in interconnecting circuits. In 1963 the firm sold it along with the ATLAS, its largest product. Competition for the sale of computers and the concurrent expense of developing the ATLAS and ORION placed a serious drain on the company's financial resources. As a result, Ferranti Ltd. sold off its computer business to International Computers and Tabulators (ICT) in September 1963. Management had concluded that computer development would require enormous outlays of money and energy at a significant sacrifice to other sectors of the business and in the face of growing competition. So, Ferranti, as did many U.S. companies during the same period, chose to stay with more traditional electronics and to abandon computers. At the time of sale, 25 percent of all computers installed in Great Britain (as measured in monetary worth) had come from Ferranti.

For further information, see: Simon Lavington, *Early British Computers* (Bedford, Mass.: Digital Press, 1980).

G

GARTNER GROUP, INC. This consulting firm has become one of the most successful in the short period of its existence as an accurate forecaster of the data processing industry's trends. It was founded in 1979 by an ex-International Business Machines Corporation (IBM)† employee, Gideon I. Gartner. Sales in 1980 reached $600,000. With average growth rates in business volume each year of between 60 and 65 percent, by 1982 sales totaled $3.2 million and in 1985, $14 million. In 1985 the firm had 140 employees.

This privately held company had nearly 1,000 clients by the mid–1980s, one-fourth of whom were users of data processing, another 125 vendors, and the remaining primarily institutional investors. Unlike International Data Group (IDG)†, another major consulting firm within the industry that focused on all aspects of data processing, the Gartner Group was more business-oriented. It tracked the successes and failures of companies within the industry, IBM's large processors and who was using them and how, and evaluated the financial worth of various organizations. In short, it studied the industry much like an investment firm on Wall Street. The period between 1983 and 1985 witnessed an expansion in Gartner's services. This firm now provided analysis of local-area communications, office information systems, personal computing, small computer systems, and strategies emerging from telecommunications firms. The focus was on what vendors were doing and why, and how users could take advantage of specific trends.

Almost from the beginning, Gartner became a highly quoted firm within the industry. When major products were announced by a large vendor, the Gartner Group would frequently be quoted on what such events meant. Forecasts about the success or failure of a firm became news items largely because this company had a high rate of accuracy. Its projection of Storage Technology Corporation's (STC) demise made in 1981 was accurate and came a year before Wall Street reached the conclusion that this high-tech company was in trouble. In late 1983

Gartner predicted that the peripherals industry would shrink rapidly, destroying such firms as Tandem Corporation and Seagate Technology, Inc. The rest of the industry did not reach the same conclusion until late 1984.

For further information, see: John Desmond, "Wall Street Style Helps Gartner Grow," *Computerworld*, June 17, 1985, pp. 91, 96, 101.

GE. See GENERAL ELECTRIC

GENERAL ELECTRIC (GE). GE is one of the oldest electronics firms in continuous operation in the United States. It was founded in 1892 as a result of a merger between Edison Electric Company and the Thomson-Houston Company. Its early years were dominated by the introduction of and sale of electrical lights and railway equipment. Yet during its first fifty years, GE became a formidable electronics company with considerable technical and financial resources capable of moving into every major field of electronics from the sale of electrical-generating equipment to consumer products. During the 1950s it did not venture into the production of computers as quickly as other machine manufacturing or electronics firms, such as Sperry Rand† or International Business Machines Corporation (IBM).† But when it did, GE's technical contributions were important, particularly in the development of time-sharing software. It played an important role as a heavy user of computers in the 1950s, as a vendor in the 1960s, and again as a leading-edge user of such technology in the 1970s and 1980s.

During the 1920s, GE introduced electrical appliances such as toasters, refrigerators, and, most important, radios. Its primary competition came from such electronics firms as Westinghouse and, for radios, from Radio Corporation of America (RCA).† It sold electrical-generating facilities throughout the United States during the 1920s and 1930s. It did considerable research into communications and played an important role in introducing products in that field throughout the 1930s and 1940s. During World War II, like many companies that would later be leaders in the data processing industry, GE did research and development work for the U.S. government. This work gave its scientists and engineers an appreciation for the benefits of intelligent machines. In 1954 GE bought the first UNIVAC* ever acquired just for commercial applications. Until this time, all computers had been sold for use in government contracts, particularly for defense applications. Thus in 1954, a milestone was reached with the purchase of a machine for commercial applications not intended for use in defense or census taking. This purchase was important because it encouraged other companies to buy computers to use for such applications as inventory control, payroll, and accounting. Over the next several years, GE installed dozens of computers, making it one of the heaviest users of such technology in the world. This experience gave GE considerable expertise as users of systems that would later allow the company to design data processing technologies with their customers in mind. The concept of "user friendly" systems dates back to GE's early experiences.

GE failed to become a major producer of computers during the 1950s because its top management did not appreciate the potential new market. The company had the technical capability to develop such machines and the financial resources required, but like many other electronics firms in the 1950s, the company had little faith in the commercial potential of computers, despite the fact that executives used them within the firm. Philip Reed, chairman of GE during the 1950s and head of the company since 1939, felt uncomfortable about the potential of a new market. Yet he encouraged the use of UNIVACs and other computers to handle administration, inventory control, and finances. Thus, by the late 1950s GE was one of the most extensive users of all forms of data processing hardware.

GE used computers particularly for accounting and corporate planning. The company learned how to cost-justify such technology as well as to appreciate the costs of maintaining and using it. GE developed methods for measuring capacity and then planning its use. Its techniques were applied by other companies throughout the 1960s and 1970s. Yet the same management that found so many uses for computers only saw a market for some 200 to 300 other companies using large mainframes and perhaps several thousand midsized processors. Reed felt that with IBM and Sperry Rand already in the market startup costs could not be justified. Instead, he opted to reduce GE's debt, which went from over $200 million in 1945 to $1.7 million in 1954. For this reason, the company did not participate in the development and marketing of the first generation of computers.

A dramatic change came in 1958 with Reed's retirement. Ralph Cordiner, GE's president for the previous eight years, took over the top position. He made a number of changes quickly, one of which was to participate in the data processing market. His goal was to use GE's vast experience as a user of computers, along with its technical staffs, to develop computer products for the commercial and scientific markets, and to seize a part of the market, especially that owned by IBM and Sperry.†

Until this time, development work in data processing had been primarily for government projects. GE, for instance, built the Office of Air Research Automatic Computer (OARAC) in 1953 and in 1956 cautiously entered the commercial market with the Electronic Recording Method of Accounting (ERMA) device, research for which had partially grown out of government work. Most of ERMA's development, however, came from an important project with the Bank of America in which the two firms developed a machine to process checks. This effort represented one of the earliest commercial projects in the data processing industry. Its significance was further enhanced by the fact that ERMA, at a cost of $60 million for the project, was the largest nongovernment contract to date involving the construction of a computer. Thirty such devices were eventually built for the Bank of America. At the same time, GE was under contract to the

National Cash Register Company (NCR)† to manufacture data processing equipment.

In the late 1950s, after Reed departed the company, GE introduced the GE 312 for process control and, in 1961, the GE 225. The GE 225 was used for both scientific and commercial applications. It also had an early variation of a COBOL* compiler called GECOM. In 1963 the company introduced a variety of data processing products including the GE 215 and the GE 235. Each offered improvements in circuitry and efficiency at lower costs over older models. In 1963 GE also announced DATANET–30, a computer that could communicate with other computers via telecommunication (telephone) lines. This major announcement for the industry meant that work could be transferred from one machine to another along with data. Thus, if one machine had too much work while another in an organization did not, work could be shared efficiently. No other vendor at that time had similar technology commercially available.

GE's work in communications extended beyond the DATANET–30 into time-sharing. GE started its own service bureau, using telecommunications from various users to GE data centers. Although the use of service bureaus and new products helped sales, in 1965, for instance, GE remained fifth among manufacturers of computers. In 1965 IBM first shipped S/360* computers in quantity, and GE beat NCR and RCA, but not IBM or Burroughs,† in sales. GE's growing experience with data processing products led management to recognize that additional millions of dollars would have to be invested for GE to stay in the computer marketplace. The company was suffering significant losses from computer sales—$100 million alone in 1966. Although the company was in good financial shape, with net income in 1966 of $355 million, stockholders increased their pressure on management to reduce and eliminate the losses.

In the 1960s informed observers believed GE had great potential as a major force within the data processing industry. GE's revenues in 1964 were $5.1 billion and, by the end of 1969, $8.4 billion, insuring the company a comfortable niche in the FORTUNE 500 family, usually in sixth place. In contrast, IBM's revenues were nearly a third less. GE's revenues from data processing during the 1960s never exceeded 3.5 percent of total sales, whereas all of IBM's revenues came from data processing goods and services. Other GE interests at the time included plastics, chemicals, motors, consumer goods, televisions, and diesel electric locomotives. Power-generating equipment and defense work continued to generate nearly half of all revenues. Throughout the 1960s, industry watchers continued to believe GE had the potential of dominating the young industry.

Yet as Franklin M. Fisher put it, "the sleeping giant never woke up." At the end of the 1960s, GE sold off its computer business to Honeywell. Historians blame GE's departure from the computer business on management. GE's executives never fully committed the company's skills and resources to the demands of the market, and, finally, they did not keep their product line up-to-date technologically in order to stay competitive.

Despite these serious charges, products continued to emerge from GE, though not with the impact required to make it a major force within the industry. Already mentioned were the 200 series which were superseded by the 400s targeted at IBM's 1401s. GE was forced to sell the 400s against IBM's S/360s which were being installed for both scientific and commercial applications. GE lowered its prices by 8 and 15 percent to counter IBM's prices for the S/360, and by 1970 announced four models of the 400s, although it never delivered all of them. In July 1964 GE brought out the 600 series to counter IBM's S/360 announcements. The GE 625 and GE 635, later also the GE 615 and GE 655, provided a variety of sizes. All of these machines had real-time capabilities for online applications and could handle time-sharing. There were problems with the operating system (GECOS) which, in combination with some hardware problems, created grave concerns for GE. Technical flaws were not resolved during the development and production phases and were only uncovered once installed in customers' data centers. In late 1966 GE felt compelled to begin retiring this entire product line. By then, GE's credibility within the industry had been severely damaged.

Losses from the increased production costs of these machines and soaring maintenance expenses were multiplied by expensive false starts on projects that were never completed. For example, the WXYZ computers in the early 1960s were never fully developed. The X machine eventually emerged as the GE 400, but the investment made in the other three machines was never recaptured. Another aborted project was the GE 100 series which was killed in 1966. Although other projects were in process in both Europe and the United States, management was never able to link developments in various countries and laboratories as IBM was able to do so well.

GE's rapid turnover in its data processing management compounded its problems throughout the 1960s. Managers were not linked to earlier projects. Thus, new programs would be started without appreciating the lessons of past experiences. Projects came and went without any continuity with the past. During this era, top management frequently had little or no experience with data processing or with manufacturing computers. Neither did it understand how to sell such devices well. These problems were publicly known, which also hurt GE's credibility. GE had failed to focus on computers with the intensity of either management concern or investment required to compete effectively against IBM, Honeywell, Burroughs, and RCA in the 1960s. Although many of its problems existed elsewhere, as for example at RCA, lack of skilled management in data processing, insufficient product development, poor quality control, and a lack of adequate research and development funding were very damaging. Furthermore, GE came into the market late, participating after the first generation of hardware had already reached maturity. Meanwhile, for those participating in the creation of the first generation of computers, many of the vital lessons had been learned about how to thrive in the market, experiences GE had to acquire late.

Thus, GE faced the second half of the 1960s with considerable exposure. Product development was scattered across five countries, and there were thirteen plants among 8,000 employees. Despite problems with computers, GE did manage to introduce additional peripheral equipment in this period and by the mid–1960s had established fifty time-sharing systems serving 100,000 customers. Its service bureau work remained intact into the 1980s.

By the late 1960s GE's computer product line, particularly the 400 series, was becoming obsolete while incompatibilities with other GE product lines were rapidly becoming unacceptable to customers. With sales dropping, the company finally decided to conduct a major review of the situation. GE formed the Ventures Task Force in late 1969 which reviewed various segments of GE's businesses, including computers. This group recommended that GE sell its computer operations in order to stem further operating losses, to lessen the demand for large amounts of capital that only drove up the cost of debt, and to eliminate the requirement to replace the entire computer line with newer technologies. The study group expressed the fear that GE would lose its customer base in other product areas. There was already little hope that its computer customers would remain loyal much longer.

The Task Force's recommendations were accepted, and so the manufacturing of computers and peripherals, along with other related segments of the business, were sold to Honeywell, which then formed a subsidiary to handle its own and GE's computer business. GE kept an interest in the new subsidiary and reserved the right to continue some data processing development. By selling off its computer business, GE avoided major losses in the future. Auditors concluded that between 1957 and September 1970 GE had suffered losses of about $163 million. Yet GE expected certain portions of its products to generate profits in future years. The sale to Honeywell also balanced these numbers.

The sale was concluded in May 1970 and called for the creation of Honeywell Information Systems (HIS). GE would own 18.5 percent of HIS and would obtain 1.5 million shares of Honeywell common stock and $110 million in the buyer's subordinated notes. The Honeywell stock then had a market value of $120 million. GE's 18.5 percent ownership was valued at $100 million by Honeywell. In 1971 GE obtained $113.2 million in additional stock in exchange for the $110 million in notes. Additional transactions between the two companies continued during the 1970s. In 1976, for example, GE traded one-third of its interest in HIS for 800,000 shares of Honeywell stock and the following year the rest of its ownership in HIS for an additional 1.4 million shares. The value of both stock transactions totaled approximately $104 million. Thus, with the time value of money and inflation considered, GE made a profit. For Honeywell, the acquisition made it the largest manufacturer of data processing equipment, second only to IBM.

Although no longer a major producer of data processing equipment, throughout the 1970s GE continued to be an important user of such technology. It became a leader in the application of technology in quality control and product developments for the consumer market. It was also a heavy user of data processing in defense projects. By the early 1980s, it was considered a major customer by many large data processing companies, including IBM.

For further information, see: Terrence E. Deal and Allan A. Kennedy, *Corporate Cultures: The Rites and Rituals of Corporate Life* (Reading, Mass.: Addison-Wesley Publishing Co., 1982); Charles and Ray Eames, *A Computer Perspective* (Cambridge, Mass.: Harvard University Press, 1973); Franklin M. Fisher et al., *IBM and the U.S. Data Processing Industry: An Economic History* (New York: Praeger Publishers, 1983); Katharine D. Fishman, *The Computer Establishment* (New York: Harper & Row, 1981); Dirk Hanson, *The New Alchemists: Silicon Valley and the Microelectronics Revolution* (Boston: Little, Brown & Co., 1982); Robert Sobel, *IBM: Colossus in Transition* (New York: Times Books, 1981).

GUIDANCE OF USERS OF INTEGRATED DATA-PROCESSING EQUIPMENT. See GUIDE

GUIDE (Guidance of Users of Integrated Data-Processing Equipment). GUIDE is made up of users of International Business Machines Corporation's (IBM's)† computers, particularly its large systems. It was established in 1956 with representatives from forty-four companies to share information informally about the use of IBM's products. In 1970 it was established more formally as a not-for-profit corporation called GUIDE International Corporation. Ten years later it had over 2,400 installations as members. An IBM S/370* Model 115 or bigger, or an IBM 4300 computer was required to qualify as a member installation in the early 1980s. Because of those requirements, its technical focus has been on large data center products from IBM involving computers, peripheral equipment, and, in recent years, such software as operating systems* and database management software.

As with SHARE† and nearly a dozen other user groups active in the data processing industry, GUIDE's goals were service-oriented. First, it sought to promote the use of professional methods of data processing management. Second, it established an extensive communication network with IBM to exchange information concerning that company's systems. This effort involved assessment of the use of products, requirements for additional functions (through task forces or GUIDE committees), and reviews of forthcoming trends in products. Its third mission was to influence public opinion concerning issues relevant to the entire data processing industry. GUIDE held two annual conferences, and by the early 1980s over 4,000 people had participated in some 150 different sessions. Particularly since the 1960s, GUIDE has profoundly influenced opinions within and outside of IBM concerning that company's products. Its educational programs have promoted the dissemination of information concerning IBM's systems. GUIDE publishes proceedings of its annual conferences along with various other monographic material.

For further information, see: GUIDE International, *GUIDE Publications Index* (Chicago: GUIDE International, 1985); T. F. O'Leary, Jr., "GUIDE," in Anthony Ralston and Edwin D. Reilly, Jr., eds., *Encyclopedia of Computer Science and Engineering* (New York: Van Nostrand Reinhold, 1983): 670–671.

H

HARRIS CORPORATION. This firm provided microelectronic and computerized technologies for use in the U.S. space programs of the 1950s and 1960s. By the 1980s it was also an important supplier of data processing products for office applications. The firm began in 1950 as Radiation, Inc. It was originally established in Melbourne, Florida, as an electronics firm by two engineers, Homer R. Denius and George Shaw. They built various space-related electronics throughout the 1950s, including mechanisms to do tracking and pulse code modulation. In 1956 the company went public with 150,000 common shares valued at $750,000. In the late 1950s Radiation provided portions of the Telestar and Courier communication satellites, parts for the Nimbus and Tiros weather satellites, and systems for use in such U.S. missiles as the Atlas, Polaris, and Minuteman by the end of the 1970s.

The 1960s represented a period of major growth and change for the company. In 1963, for example, it opened a microelectronics plant to build integrated circuits. These were used for a variety of products in digital and space communications, data management, and computer-based control systems, and almost all the customers were within the federal government. It built additional satellite tracking systems and became a major supplier of high-technology electronics for space projects by the middle of the decade. Radiation ended 1966 with sales of over $50 million and with 3,000 employees. The next year it merged with the Intertype Company which manufactured printing equipment. Through that merger the company achieved sales of $250 million and had 12,000 employees. Business continued to prosper, and in 1969 the company turned in sales of nearly $340 million. That year it expanded yet again by acquiring RF Communications, Inc., of Rochester, New York, which manufactured two-way radio products. The company moved deeper into the world of data processing by building computer-controlled data-handling systems for preflight checkout in Apollo Lunar Spacecraft and won a contract to build a digital command/control system for the Gemini missions. In 1969 the firm also began selling Fototronic-CRT typesetters which were in effect computerized terminals. More computerized technologies appeared in printing products as well.

In 1971 the company acquired partial interest in Datacraft Corporation which made computer systems similar in size to the PDP 11 and International Business Machines Corporation's (IBM's)† S/360* Model 50, in addition to core memories usable either on IBM's computers or its own. The next year it bought CSI, a subsidiary of UCC, later known as the Wyly Corporation. CSI made programmable communications controllers compatible with IBM's computers. It also produced terminals, printers, and card readers. Then in 1974 the company changed its name from Harris Intertype to Harris Corporation, the name it currently uses. From 1976 through 1979 Harris brought out a variety of computer-related products which included computers, front-end processors (such as control units), and terminals. It had become a vertical operation with products for distributed data processing, computers, data communications, office systems, text editing and composition (printing), control, and satellite communications among its main areas of specialization. Harris also had its two-way radio business, had begun manufacturing television and radio broadcasting equipment, and, of course, had its integrated circuits operation. It continued to be an important supplier of specialized electronics for the government. Thus, like many of the early companies in data processing, it was born out of contracts with the federal government, flourished in its early years because of that relationship, and, while exploring commercial opportunities, maintained close links with the original source of business.

The products Harris made in the field of data processing allowed the firm to sell to a broader set of customers. Thus, it enhanced its 1600 distributed data processing products in 1979 with a bigger computer that had twice the memory of the old 1600. In 1978 Harris had already introduced the Series 500 made up of the 550 and 570 systems, disk-oriented computers with virtual memory for use by scientific, industrial, and educational end users. These systems were typical of the period in that they handled concurrent time-sharing, interactive computing, batch and remote job processing, and real-time transactions. In 1979 Harris enhanced this family of computers with the Series 800, a large computer that supported up to 128 interactive terminals at the same time. The 850 processor could be configured with up to 3 million bytes of memory, making it a large intermediate-sized computer.

The 1980s began with yet another acquisition, this time the Farinon Corporation of San Mateo, California, a firm that specialized in the manufacture of telecommunications systems, microwave radios, digital telephone switches, and other telephonic devices. As the company moved deeper into telecommunications, data processing, and office systems, it decided to shed its printing products and, thus in April 1983, sold off this end of its business, making it now totally an electronics firm. Then in October, Lanier Business Products, Inc., merged with Harris. Lanier, headquartered in Atlanta, Georgia, was by then an important American manufacturer of office automation equipment. This merger therefore placed Harris squarely in the middle of the office automation market of the early 1980s along with IBM, Wang, and others. It could now offer word processing systems, business computers, dictating equipment, and copiers, and had access to Lanier's nationwide sales and distribution network.

In 1984 sales amounted to $2.19 billion for the company as a whole, and it employed 30,000 people. It had thirty-five plants scattered throughout the United States, Canada, Asia, and Europe. *Datamation*, a major journal in the data processing industry, ranked Harris thirty-seventh among the top 100 data processing companies in size as measured by sales. The industry as a whole noted that this company, with its image of scientists and engineers, had been working hard in the early 1980s to become a marketing organization. Data processing revenues alone that year rose 26 percent to some $730 million, with nearly half of these still coming from the federal government. This suggests that marketing remained an issue to be worked on. In early 1985, owing to its inability to compete profitably against other computer vendors with terminals, Harris stopped marketing such products. Computer sales, however, kept growing, and new products were announced. It also introduced a low-end departmental system called the H–60 and HarrisNet which was the company's local area network (LAN) system. It even offered the Harris PC, a small desktop computer compatible with IBM's office products. In the early 1980s it continued to make integrated circuits and had a major contract with Cray Research to provide components.

For further information, see: Pamela Archbold and John Verity, "The DATAMATION 100," *Datamation*, June 1, 1985, passim; Franklin M. Fisher et al., *IBM and the U.S. Data Processing Industry: An Economic History* (New York: Praeger Publishers, 1983).

HEWLETT-PACKARD COMPANY (H-P). In its first thirty years of existence, Hewlett-Packard sold measuring instruments. Beginning in the mid–1960s, it turned to data processing products, and by the late 1970s had become an important manufacturer of minicomputers and hand-held calculators. In the early 1980s surveys by *Fortune* Magazine rated H-P as one of the most admired corporations in America. Furthermore, its unique management style was imitated by many high-technology companies operating in California and in Massachusetts—two important locations for young, small data processing companies.

In 1938 William (Bill) Hewlett and Dave Packard,** engineering graduates of Stanford University and friends for years, developed an audio oscillator in the garage located behind the home of Packard's parents in Palo Alto, California, which worked with a broader set of conditions than existing devices and cost $55 instead of the $500 which was then common. Thus H-P was born. Coincidentally, nearly four decades later, Apple Computer,† manufacturers of some of the earliest desktop computers, was also born in a California garage. The oscillator was employed to test sound equipment and was first introduced at a meeting of the Institute of Radio Engineers (today called the IEEE*) as the Model 200A. Although few orders were taken, Walt Disney Studios wanted a variant (which they built, called the Model 200B). Early in 1939 the two engineers formalized their partnership. They began to create other instruments such as a voltmeter built with vacuum tubes, and in 1942 the firm constructed its first building. The need for military electronics during World War II provided additional growth opportunities. In 1946 H-P brought out a microwave signal generator, and the following year it formally incorporated.

In 1950 H-P had sales of $2 million, 200 employees, and 70 products. Throughout the 1950s the product line broadened to include a high-speed electronic counter, a calibrated laboratory oscilloscope, and replacements for earlier offerings. In 1958 H-P bought the F. L. Moseley Company of Pasadena, California, a little firm that built high-quality graphic recorders. In the early 1960s H-P acquired the Sanborn Company of Waltham, Massachusetts, an early manufacturer of electrocardiographic equipment and supplier of recording instrumentation. A second acquisition involved the F&M Scientific Corporation of Avondale, Pennsylvania, which made gas chromatographs. This acquisition allowed H-P to apply its solid expertise in electronics to the fields of medicine and analytical chemistry. In 1958 it already boasted a product line of 373 electronic test and measuring instruments and parts, and then in 1959 it established a marketing organization in Europe. Despite these acquisitions and expansions in both products and markets, customers throughout the 1940s and 1950s were essentially the same—scientists and engineers. This circumstance did not change essentially until the late 1960s when the company began marketing data processing products.

An important step toward the world of data processing was taken in 1961 when H-P created an organization to develop solid-state components in Palo Alto. This action reinforced H-P's presence and importance in that part of California that would later be known as Silicon Valley, home for hundreds of companies manufacturing and developing various data processing technologies. In the 1970s in particular, many organizers of such firms would come out of jobs originally held at H-P and looked back to those days as exciting and intellectually rewarding. By the middle of the 1960s H-P had acquired various sales organizations that earlier had sold its products. By then H-P was continuing to introduce electronic instruments that could produce measurement data rapidly, almost computer-like. Then, in November 1966 the company announced its first processor, called the Instrumentation Computer, which was intended to work with earlier instruments to provide computational support. The company was successful with its initial forays into data processing, encouraging it to continue. In 1972, for example, revenues from data processing products comprised 15 percent of the company's performance, or some $68 million out of nearly $480. In 1979 revenues from data processing products exceeded $1 billion or nearly

50 percent of H-P's total income, making it a major company within the data processing industry.

With the hand-held calculator so common today, it is difficult to believe that there was ever a time when everyone did not have at least one. One of the earliest and best of these calculators came from H-P. Announced in January 1972, the "electronic slide rule" came out as the HP–35 scientific calculator. As its nickname implied, the device was intended to replace the slide rule,* long the favorite portable calculator of generations of H-P's customers—engineers. Between this product and similar ones from Texas Instruments,† the slide rule passed into history almost overnight. Keuffel and Esser, the leading U.S. manufacturer of slide rules, killed the product in 1973 largely because of the arrival of hand-held calculators. The H-P calculator was the best in the market as well as one of the most expensive. H-P also introduced the first programmable hand-held calculators and even in the 1980s still produced high-function devices of this type. The HP–35 grew out of the company's earliest experiences with computational equipment, especially the desktop calculator introduced in 1968 to work in conjunction with the instrumentation product line of that era.

In early 1972 H-P also brought out another product important to the history of data processing called the HP 3000. It was a minicomputer system that performed general-purpose computation. It could do time-sharing, multiprogramming, batch, or online processing, and it supported various languages found on large mainframes. Following initial problems with performance, which were fixed by 1976, a new version appeared called the 3000 series II. In 1978 the series III and then the series 33 came out and in 1979, the 3000 series 30. These later devices could be configured with up to 2 million bytes of storage, communicated with others, and supported remote teleprocessing. H-P added an operating system and software to support as many as sixty-three terminals per configuration. Compilers came out to handle programs written in COBOL,* FORTRAN,* and BASIC,* and the company developed IMAGE, a database management system.* H-P also manufactured peripheral equipment for these systems, including among many products terminals, disk drives, and printers.

The HP 3000 played a critical role in the company's evolution during the late 1970s. They were sold successfully for such applications as order entry, production control, and warehousing. Software appeared for the textile industry and manufacturing shop floor applications. Advertising firms bought the equipment, as did engineers. The greatest successes of these minicomputers, however, were in manufacturing plants. There H-P competed against Sperry,† International Business Machines Corporation (IBM),† and other major computer vendors in the late 1970s and early 1980s.

In 1984 total revenues reached $6.3 billion, of which $3.4 billion came from data processing products, a growth in data processing revenues of 36.2 percent over 1983. In 1984 it was the eighth largest company in the industry. It employed 80,000 people in eighty countries and boasted a product line of 9,700 different items related to data processing.

To a large extent, the company's success seemed to rest on its high-quality products and on its loyal base of scientific and engineering customers. In the early 1980s the company elected to enter the office automation market, hoping to build on its install base of some 17,000 3000 series minicomputer customers. It introduced new products including desktop computers (to compete, for example, against IBM's PC and Apple's products). In 1984 H-P enhanced the 3000 with the series 37 which it targeted at commercial end users while it continued marketing the HP 3000 series III. By 1985 it could claim to continue holding the number one position as supplier of analytical instrumentation. It had been second in the manufacture of minicomputers for a number of years. In 1984 microcomputers accounted for less than 5 percent of total sales.

Another distinguishing feature of H-P is its management style which has influenced that of many data processing firms, especially the companies in Silicon Valley. In effect, it is a style that takes great pains to indicate to employees that top management is concerned for their welfare and appreciates people. This pattern of behavior is reflected in such things as stock options, a large collection of employee benefits, and personal contact with people at all levels of the company by executives and management alike. In some locations it means not wearing coats, suits, or ties while company picnics and parties have become common. Such firms as Tandem Computer and Apple publicly state that they want their organizations to operate in a similar manner. Ex-managers and engineers from H-P in other firms have also exported the corporate culture to other institutions.

For further information, see: Franklin M. Fisher et al., *IBM and the U.S. Data Processing Industry: An Economic History* (New York: Praeger Publishers, 1983); Hewlett-Packard Company, *Hewlett-Packard: A Company History* (Palo Alto, Calif.: Hewlett-Packard Co., 1983); Everett M. Rogers and Judith K. Larsen, *Silicon Valley Fever: Growth of High-Technology Culture* (New York: Basic Books, 1984).

HOMEBREW COMPUTER CLUB. This informal club was composed of people in California interested in microcomputers and sometimes of counterculture notions during the mid- to late 1970s. From its ranks of engineers and computer enthusiasts came the founders of many data processing companies in Silicon Valley, that part of California that became home to hundreds of small data processing firms. Members included the entrepreneurs who established a variety of microcomputer companies—for example, Adam Osborne** and Steve Jobs** of Apple Computer.† The club provided an early impetus for the development and then expansion of the data processing industry in California. Because of that role, it can be considered one of the most important data processing organizations of recent years.

With the introduction of the Altair microcomputer in 1975, the immediate need for users to form a group became evident. Until then those interested in personal computing had essentially worked on their own in isolation, but now there was a specific reason to band together. From the beginning, the club attracted counterculture engineers, computer specialists living in California in the late 1960s and early 1970s, and frustrated yet-to-be-successful high-tech entrepreneurs, most of whom were under the age of thirty. In 1975 a group of people interested in Altair or who were building their own small computers met under the name of the Amateur Computer Users Group but quickly acquired the name of Homebrew Computer Club. They held their first meeting on March 5, 1975, in the garage of one of its members, Gordon French. Thirty-two people attended this first gathering, but by the third meeting, several hundred were involved. Early discussions concerned the Altair and other technical topics such as which chip* to use as the standard for making all other micros, how to improve memory, and how to design more efficient logics for such micros. Very quickly the casual, anarchic characteristic of the club was set by its members, particularly by Lee Felsenstein, its dominant member and the future designer of the Osborne 1 microcomputer. The club had no hierarchy of officers; Felsenstein was simply master of ceremonies. Within a year of its formation, the club's sessions had to be moved to the auditorium of the Stanford University Linear Accelerator Center where nearly 750 people now attended meetings. Information on members is sketchy inasmuch as the organization had no dues or formal membership, and was always open to anyone interested in computing. However, it did publish a newsletter that reflected the concerns discussed at the meetings.

Sessions were attended by Adam Osborne who sold copies of his books out of cardboard boxes, and by Jobs and Steve Wozniak,* both of whom later formed Apple Computer, to mention a few important people. Branches of the club were established in San Francisco and at Berkeley. At the meetings people exchanged information about all aspects of microcomputing's technology, while some designed such devices and sold them.

The club also fostered a counterculture within data processing. In the mid–1970s the data processing establishment consisted of users of large computers and such vendors as International Business Machines Corporation (IBM)† and Honeywell. Homebrew's members did not identify with corporate America, let alone with the mainstream of its technologies. This is an important point because much of the research and development work being done at major institutions concerned large-scale computing, for both hardware and software, while the focus of Homebrew's members was not on either of these fields. Homebrew's concerns became critical to the success of larger institutions in the 1980s. The interests of the club's members also extended to new software and peripherals for micros and later even to large systems (particularly for storage devices such as disk drives), and to entrepreneurial exercises that ultimately led to the formation of many small firms in Silicon Valley made up of several dozen employees or less. Members' ideas received honest and critical examination at the meetings; support or disapproval at these sessions determined the fate of many small firms.

One author, Stan Augarten, has described the club as a "joyous anarchy." In encouraging the use of new technologies and in sharing new or different ideas, it had the same role that personal computer (PC) clubs did in the 1980s. Some of the better known firms that emerged from this group, in addition to Apple and Osborne, were Processor Technology which built better memories for the Altair micros than Altair did at the time; Cromemco; North Star; Vector Graphic; and Godbout, all of which had been founded by 1977. With the growth of micros in 1977, user groups within Homebrew began to emerge, for example, to support users of micros made by Apple and Altair or to support the views of Processor Technology.

By the end of 1979, Homebrew had been eclipsed by rapid changes within the industry. Microcomputing was no longer a small, almost countercultural event with members advocating "computing power for the people." The "people" were acquiring it from large firms such as Apple and Texas Instruments.† Then in 1981 IBM legitimized microcomputing when it announced its PC. From then on corporate America had made microcomputing an "establishment" part of the industry. Also by 1980, the real frontier had begun to shift dramatically away from the development of new chips and microprocessors by garage-size companies to large firms. The leading-edge for small developers was now in software. But in its day, Homebrew had spawned a revolution in microprocessing, serving as midwife at the birth of personal computing for people not conversant with data processing.

For further information, see: Stan Augarten, *Bit by Bit: An Illustrated History of Computers* (New York: Ticknor & Fields, 1984); Paul Freiberger and Michael Swaine, *Fire in the Valley: The Making of the Personal Computer* (Berkeley, Calif.: Osborne/McGraw-Hill, 1984).

HONEYWELL, INC. Honeywell has had a long history as an important vendor in the data processing industry. Unlike most firms in the industry, Honeywell has traditionally been a vertically integrated supplier; it manufactures, sells, and services a complete line of products from computers to peripherals, and from operating systems* to programming languages.* During the 1960s Honeywell was a major competitor of the International Business Machines Corporation (IBM),† but by the 1970s it had settled into a position where it owned approximately 10 percent of the large computer market.

Honeywell's origins date back to the late nineteenth century when Albert Butz, a Swiss immigrant to the United States, designed the "damper flapper," which manipulated dampers in coal furnaces as a means of maintaining even room temperatures. Its second purpose was to prevent stoves from becoming so hot (due to overstocking) that they might burn down a house. The flapper was a product of the late 1800s, but in the 1920s Mark Honeywell and W. R. Sweatt, who each had their own heat control companies, merged together to sell the device. Their company, established in 1927, was called the Minneapolis Honeywell Heat Regulator Company. Over the next several decades they expanded their product line to include other heat regulating devices and various temperature control systems—products that are still being manufactured and developed at Honeywell and which generated nearly 60 percent of the company's revenues in the early 1980s. In 1934 the firm acquired Brown Instrument Company of Philadelphia. At the time Brown was a major vendor of recording, indicating, and controlling instruments in the United States. Thereafter Honeywell began more frequently to use electrical components in its products. During World War II Honeywell, like most American companies, manufactured items needed by the military—in particular, automatic instruments, such as the C–1 autopilot system.

Honeywell's growing skills and its participation in the electronics field were preconditions for its entrance in the 1950s into what would develop into the data processing industry. Its expertise allowed the firm to appreciate the technical significance of Bell Laboratories',† development of the transistor at the end of the 1940s and to understand its market potential. The company soon became interested in computers and, in 1955, joined with Raytheon Company† in a joint venture to construct such devices. They, in turn, formed the Datamatic Corporation, which designed and built large-scale computers for commercial users. The initial computer was based on work done earlier by Raytheon in its construction of the RAYDAC (Raytheon Digital Automatic Computer).* In short, Raytheon brought to the venture an extensive background in computers at a time when such expertise was rare. The acquisition was a logical step for Honeywell, which, at the time, was one of the largest manufacturers of automatic control equipment in the United States and was also interested in expanding its opportunities.

Raytheon owned 40 percent of Datamatic, Honeywell, the other 60 percent. In 1955 Honeywell's sales reached $244 million with a net income before taxes of $40 million. Its assets were valued at $164 million. Raytheon's sales were $182 million with before-tax net earnings of $9 million and assets of $82 million. Put another way, the two were about the same size as IBM or Sperry Rand Corporation†, the two other giants in the emerging computer industry. Datamatic's first product (actually, Honeywell's first computer-based offering) was the D–1000, which appeared in 1957. It was a large computer that sold for nearly $2 million. Honeywell built the computer and its tape drives but acquired other peripheral equipment from various vendors. These included card punch equipment from IBM and printers from Analex. The company sold between eight and ten copies of the system. In 1957 Honeywell bought out Raytheon's share of Datamatic; it was a move that gave Honeywell a strong position in the middle of the new industry.

The following year Honeywell introduced its second computer, a processor based on transistorized technology dubbed the Honeywell 800. It was the firm's first medium-sized processor. In 1959 it announced the H–290 which, as a continuous-control processor, was a machine designed for use by public utilities. It was sold to chemical, petroleum, and other energy-related companies. In December 1960 Honeywell announced the H–400, a computer fully compatible with the earlier 800 at half the price. Initial shipments of the machine came in late 1961. Looking at the offerings as they existed in 1961, Honeywell had a broad range of products from small to large computers. That same year it began to introduce software products, the most important of which was FACT (fully automatic compiling technique).* A programming language for business applications, FACT had been designed to compete against IBM's COMTRAN and the various compilers for COBOL* appearing in the market. It failed and was replaced by a COBOL compiler.

Like other successful firms in the industry, the company recognized the need to continue introducing more products rapidly. In 1962 it announced the H–1800, a computer that could be used for both commercial and scientific computing. It also negotiated an arrangement with Nippon Electric Company whereby the Japanese firm would manufacture and sell Honeywell's products in the Far East. Reflecting newer technologies and better price performance, Honeywell brought out the 1400 in 1963 as a replacement for the 400 series. In December 1963 it introduced the H–200 as a means of taking away from IBM customers who used the IBM 1400 (not to be confused with the H–1400).

The upshot of having introduced all these products was Honeywell's growing presence in the computer industry. In 1958 its revenue from data processing products sold in the United States had been $1 million. In 1963 it reached $27 million. In that last year data processing revenues accounted for 5 percent of the company's total sales. Thus while it was an important firm in the young industry, the company had not yet made the full commitment to computers that such other firms as IBM and Sperry Rand had, a factor that caused many companies, including Honeywell, to be eclipsed in the market in future years by IBM.

In April 1964 IBM announced its S/360.* At the time, it was clearly the most important new product in the history of the data processing industry, and, by the late 1960s, in American history. In short, it made IBM a very large corporation. For the rest of the industry, it ushered in a new era of complexity, technological advances, and severe competitive pressures for all. Many companies came and went over the next fifteen years and at a more rapid pace than had been evident before. Those that were prepared to make a life or death commitment of resources to play in this market stood a better chance of survival than those that only dabbled in computers. Those that were not committed usually waned and in many cases made sensational withdrawals from the market. Honeywell, like other competitors of IBM, reacted to the S/360 by making all of its computers compatible in subsequent years, as IBM had done, while lowering purchase prices on older models. It announced new computers called the 120, the 1200, the 2200, the 4200, and the 8200, which were of various sizes but all compatible with the H–200.

S/360 launched what many historians have dubbed the computer wars of the 1960s, in which major vendors battled for market shares and positions. In fact, the wars had started earlier, around 1961 or 1962, though they heated up with the rise of the S/360. The greatest casualties were General Electric (GE)† and the Radio Corporation of America (RCA),† both of which retreated from the market. Neither was prepared to make the kind of financial commitments required to compete properly. Honeywell, still a major supplier of temperature control systems, remained in the battle. It began this round by marketing its 200 to displace IBM's 1400s—both had been made obsolete with the introduction of the S/360. The 200 proved to be a successful product, with several hundred orders logged by the company. Honeywell's sales force, which had equalled RCA's in size at the start of the decade, had grown to between 50 and 75 percent larger in number by 1965 when S/360s were shipped in quantity. By the end of 1964, the 200 had displaced some three hundred computers from IBM and was threatening to supplant nearly one thousand others. Throughout the decade, as IBM enhanced the S/360, so too did Honeywell improve its line with 200-compatible products.

Despite the company's success in attacking IBM and in bringing out new products, all was not perfect. Its 8200 computer, announced in 1965, was not able to attract large customers that would use the machine other than as a replacement for Honeywell's older 800 series. Its peripheral equipment was compatible with Honeywell processors and, by the end of the decade, was almost all made by Honeywell's plants. The company invested large sums in the development of the 8200 but only shipped about forty copies. In short, the company lost money on the project. This experience clearly suggested a pattern that has existed into the 1980s, whereby Honeywell's successes primarily have been within its own existing base of customers. Customers would replace Honeywell computers with newer Honeywell models. While in the 1960s and 1970s a Honeywell salesman would occasionally displace perhaps an IBM processor, such was a rare event.

The company adopted an expansionist strategy that included acquisitions. In 1966 it acquired the Computer Control Company, a leading developer of small high-performance products, and renamed it the Computer Control Division. Its products were the DDP–116, DDP–416, and DEP–516 computers, along with memory* systems. These small machines were sold to engineers and used for both scientific and electronic communications switching applications. Skills acquired with the company appeared in other products, which incorporated time-sharing, communications, and applications in the medical field. In 1969 new products began to replace these aging ones. The first was the Honeywell 316, a minicomputer that was also a digital general-purpose processor complete with a full line of peripherals. The company also created the Information Services Division in the late 1960s with sixteen data centers offering time-sharing. These centers used a new processor called the 1648 computer, which had been introduced to compete against IBM's 360 models 25 and 30 processors and Digital Equipment Corporation's (DEC)* PDP–10—all medium-sized machines.

A major reorganization in 1968 led to the creation of the Computer and Communications Group, formed out of the EDP (Electronic Data Processing) Division and the Computer Control Division. This group consolidated the company's data processing activities and became the basis of a new organization called Honeywell Information Systems, which was created in 1970 when Honeywell merged its own data processing organization with that of GE's when the latter bowed out of the market. The merger initiated a new era at Honeywell. But looking back prior to this watershed event, sales within Honeywell in the 1960s had already made the firm a sizable enterprise in the market. From domestic data processing revenue totaling $27 million in 1963, sales had climbed to $210.8 million in 1969. This sevenfold increase had taken place at a steady pace. In fact they had outpaced the great expansion of the industry as a whole and gave management confidence in believing that the decade of the 1970s would bring a repeat of previous success. The company had also invested over half of its R & D dollars into new products throughout this same period, and not all of the results had appeared yet as products.

The acquisition of GE's data processing operations in 1970 was a major event not only for the company but also for the entire data processing industry. It obviously made Honeywell a larger vendor. It also caused GE's customers to ask questions about what support and services they could expect for their machines and what plans there were for replacement products. The latter question was critical, because if compatibility was not preserved from generation to generation, migration to new systems would force the conversion of all programs, an expense that had historically exceeded the actual purchase price of computer systems themselves. The merger involved the establishment of Honeywell Information Systems (HIS) as a subsidiary of Honeywell Incorporated that would serve both sets of customers. Revenues in 1970 from the combined computer operations of GE and Honeywell reached $859 million. This total equalled about 13 percent of all the revenues of GE and Honeywell combined.

In February 1971 Honeywell announced its first computer products for the ex-GE customers called the 6000 series. This was a collection of six computers that surpassed the GE 600 line in performance and low prices. New disk drives and other peripherals were also introduced. The new processors used GE's operating system (GECOS); this satisfied the concern regarding compatibility, at least for a while. The introduction proved successful, and between 1972 and 1973 about forty-five systems were shipped to new and existing customers.

Then in January 1972 another line of processors called the 2000 series appeared, intending to replace Honeywell's 200s. The company still aimed blows at IBM, and now also at Burroughs Corporation,† with its 2020 and 2030 models. At the same time the company introduced the DATANET 2000, a programmable telecommunications processor that could be used to help manage growing telecommunication networks of remote job entry (RJE) and interactive terminals. For those familiar with IBM's widely used 3705 controller, the 2000 served the same purpose. These various products proved successful in maintaining and enhancing Honeywell's position within the industry. Revenues from HIS increased from $856 million in 1975 to $1.5 billion in 1979.

However, not all of this growth came just from the sale of 2000 series computers. As in the 1960s, part of the growth was generated by acquisitions. In addition to the GE merger, Honeywell made a similar arrangement with Xerox Corporation† when the latter retreated from the computer wars. Honeywell made peripheral equipment jointly with Control Data Corporation (CDC),† acquired Bull Company† of France, bought out GE's remaining interest in HIS, and expanded sales in Europe and into Australia. The company also elected to develop products for the distributed processing marketplace. These included several computers which formed the Distributed Processing series (DPS): the DPS/C in 1978, the DPS 8 in 1979, the DPS 88 in 1982, a variety of DPS 88 systems in 1984, and the DPS 90 in 1985. Beginning in 1976, it began to encourage Xerox's old customers to start migrating to HIS products.

By the early 1980s Honeywell had a significant niche in the data processing market. In 1982, when it ranked eighth in total revenues, its sales reached $1.684 billion. The following year Honeywell's ranking dropped to ninth, with revenues of $1.666 billion for data processing products. The corporation as a whole enjoyed revenues of $5.753 billion in 1983.

The company had established a semiconductor operation in the early 1980s that had hurt overall earnings in 1982 and in 1983. The acquisition, called Synertek, relied in large part on sales of components to the electronic games business, and in particular Atari. Since this portion of the data processing industry had been in decline, this decline was mirrored in lagging sales of semiconductor parts. But HIS now made up about 30 percent of the entire corporation, and its business volumes were healthy during this same period. New products continued to appear as well. For example, in addition to the DPS series, the 6/10 appeared in 1982 as an end user computing system complete with terminals. The large mainframe market remained a difficult one, as in the early 1980s most mainframe manufacturers were pursued by competition from Japanese vendors and IBM. The data processing industry as a whole experienced difficult years in both 1985 and 1986. This negatively influenced Honeywell's overall performance and, more specifically, that of its data processing operations.

For further information, see: Franklin M. Fisher et al., *IBM and the U.S. Data Processing Industry: An Economic History* (New York: Praeger Publishers, 1983); Katharine Davis Fishman, *The Computer Establishment* (New York: Harper & Row, 1981); Tom Henkel, "Research, Acquisitions Tell Honeywell's Tale," *Computerworld*, September 9, 1985, pp. 95, 98–99.

H-P. See HEWLETT-PACKARD COMPANY

I

IBM. See INTERNATIONAL BUSINESS MACHINES CORPORATION

IDC. See INTERNATIONAL DATA CORPORATION

IEEE. See INSTITUTE OF ELECTRICAL AND ELECTRONICS ENGINEERS, INC.

IFIP. See INTERNATIONAL FEDERATION FOR INFORMATION PROCESSING

INFORMATICS, INC. Informatics was one of the first companies formed within the data processing industry to sell primarily software and programming services. Its most notable software product was the Mark IV, a successful database management* package. The need for firms to develop products or to do contract programming became particularly obvious after International Business Machines Corporation (IBM)† announced the S/360* in 1964, creating rapid growth in the demand for computing within many organizations that did not have sufficient staffs to write software rapidly enough. During these years the data processing industry as a whole grew at over 25 percent annually, creating growth markets in a number of areas, including software; Informatics hoped to capitalize on such opportunities.

The first company, known as Informatics, Inc., was established in March 1962 and on February 28, 1974, merged with Equimatics, Inc., which was owned by the Equitable Life Assurance Society of the United States. The new entity was then named Informatics, Inc. Through the sale of stock in 1979 and 1980, Equitable divested itself of the company. In May 1982 the software house was renamed Informatics General Corporation, the name it retains today.

To understand the opportunity that led to the birth of Informatics and the environment it operated in as part of the software subindustry within data processing as a whole, a few statistics will be helpful. In 1965 the United States had only about forty to fifty independent suppliers of software. By the end of the decade, this subset of the industry had exploded with exponential growth each year. In 1968, for example, the number of software vendors approached 2,800, and the following year contract programming services constituted a $600 million business; software products were already generating another $20 to $25 million in the United States alone. In addition to these impressive sums, the total amount spent on software in the U.S. economy was far greater, going from some $200 million in 1960 to nearly $4 billion in 1965, to $8 billion in 1970, to $12 billion in 1975, and with similar growth patterns evident into the 1980s. About 90 percent of these funds went for the salaries of programmers employed by users to write in-house software. This was the background against which Informatics operated.

The original purpose of the company was to offer programming services to federal government agencies. By the end of 1969 Informatics had added other services and products, including programming for computer manufacturers and private industry; written and sold software products; and offered other services to data centers. Informatics closed out 1969 with U.S. revenues of $19.8 million. A major contributor to its revenues in the 1960s and 1970s was a software package called the Mark IV, which was first shipped in 1967. It was one of the first commercially available database management packages to function on IBM's computers. When compared to the products of other firms, the Mark IV did well, with over sixty licenses to its credit installed by September 1968. The following year that number jumped to 171; new releases appeared in 1973 for the IBM S/370* and UNIVAC Series 70 computers. Informatics opened its first two data centers in 1969 (in Los Angeles and San Francisco) to provide time-sharing and access the the Mark IV.

The company continued to experience steady growth in sales and income throughout the 1970s. Revenues reached $38.9 million in 1975, $112.3 million in 1979, and $170.1 million in 1982. During the early to mid–1980s, it divested itself of unprofitable nonsoftware ventures, including selling off its data services, and expanded into Europe. In 1984 the company reorganized into two units as opposed to eight the year before and twenty-two even earlier. One of the co-founders of the company, Walter F. Bauer, replaced Bruce T. Coleman as president and chief executive officer in 1984. In that same year $191.1 million in revenues were reported. By the standards of the data processing industry, Informatics was one of the oldest and largest software and programming firms. Moreover, unlike several thousand other firms that came and went, it has survived and usually very profitably so, with margins in excess of 12 percent in many years.

For further information, see: Richard L. Forman, *Fulfilling the Computer's Promise: The History of Informatics, 1962–1982* (Woodland Hills, Calif.: Informatics General Corporation, 1985).

INSTITUTE OF ELECTRICAL AND ELECTRONICS ENGINEERS, INC. (IEEE). The IEEE is the largest association of engineers in the world, claiming a membership in excess of 250,000 in 120 countries in 1985. It was founded in 1884 with primary focus on the educational needs of those working with electricity. During the twentieth century it has published a wide variety of journals, newsletters, transactions, and proceedings concerning electrical engineering. Yet from the inception of digital and analog computing in the 1930s, its members played important roles in the development of computing technologies. Beginning in the late 1940s, the IEEE paid attention to the subject of computers. By the late 1960s, it had become a major lobbying association within the industry and, like the Association for Computing Machinery (ACM),† an important source of good technical information on data processing through its large series of publications, seminars, and conventions.

The IEEE's primary mission is to advance the theory and practice of the electrical, electronic, and computer sciences. Initially, much of its attention was lavished on the role of electricity in such fields as lighting and later radio. After World War II it became interested in television and as that field grew in diversity, in all aspects of electrical engineering. In 1884, when it was founded in New York, it was called the American Institute of Electrical Engineers. In 1912 an organization dedicated to the advancement of technologies used in radios was formed, called the Institute of Radio Engineers (IRE). The two merged in 1963 to form the IEEE (often pronounced I Triple E).

Another organization within the IEEE, called the IEEE Computer Society (IEEE-CS), was founded in October 1951 as the Computer Group of IRE and its name was changed to the current one in 1972. The IEEE-CS, formed to exchange information solely about computers, has experienced phenomenal growth. In 1975 its membership exceeded 21,000 out of a total IEEE population of 160,000. Subsequently, the percentage of total membership in the IEEE interested in computers has gone up. The computer group has always supported its own organization and produces a series of transactions and other publications concerning all aspects of computer technology.

For further information, see: IEEE, *A Centennial Guide to Electrical Engineering History for the IEEE* (New York: Center for the History of Electrical Engineering, 1983).

INSTRUMENT SOCIETY OF AMERICA. See ISA

INTEL CORPORATION. Intel was one of the leading manufacturers of semiconductors in the United States during the 1970s and early 1980s. It was founded in 1968 and rapidly became a major supplier of large-scale integrated circuits. Within several years of its founding, it produced microprocessors and entire memory systems for computers built by International Business Machines Corporation (IBM)† and Digital Equipment Corporation (DEC)†. It marketed memory as add-on storage for computers made by these two firms in competition with them throughout the 1970s. Within ten years of its establishment, Intel's products were also buried in Wang's word processors, within Hazeltine terminals, and even in Aydin video graphics terminals. In the 1980s IBM also acquired some of its microprocessors. To support these products the company also sold software, programming languages,* and other chips* to manage such items as floppy disk drives, cathode ray terminal (CRT) displays, and communications equipment. It built microprocessors that could support such languages as FORTRAN 77,* BASIC,* PL/M, and PASCAL.

But the Intel product that gained the greatest amount of attention within the data processing industry was its add-on memory which it sold for use on IBM System/370* computers beginning in 1971. By the end of 1975 its salesmen were calling directly on IBM's customers and on leasing companies that owned IBM computers. It did the same thing for users of the PDP 11 (made by DEC) and later for the LSI–11 family, also from DEC. In 1979 the company announced the FAST–3805 disk storage subsystem, a system that proved far faster in operation than most contemporary storage systems. It was aimed at IBM's 3830/3350 and 2835/2305 customer base because these were compatible with IBM's equipment in the sense that they emulated IBM's products. The company also claimed that they operated much faster than IBM's.

In 1979 Intel also acquired the MRI Systems Corporation which gave it the capability of selling a database management system* for use on IBM, Control Data Corporation (CDC),† and UNIVAC* computers. Called the System 2000, it relied on Intel's FAST–3805 disk to take full advantage of its features. IBM acquired 12 percent of the company and later increased its ownership to 18 percent as a means of obtaining a supply of semiconductors. In the 1980s Intel made the very popular and widely used 8088 microprocessor, the basic building block of IBM's early Personal Computers (PC). Throughout its history Intel has enjoyed a reputation for producing high quality chips* and, unlike many of its rivals, has survived. Most of its competitors survived only three to five years.

For further information, see: Franklin M. Fisher et al., *IBM and the U.S. Data Processing Industry: An Economic History* (New York: Praeger Publishers, 1983); Glynnis Thompson Kaye, ed., *A Revolution in Progress. A History of Intel to Date* (Santa Clara, Calif.: Intel Corporation, 1984); Everett M. Rogers and Judith K. Larsen, *Silicon Valley Fever: Growth of High-Technology Culture* (New York: Basic Books, 1984).

INTERNATIONAL BUSINESS MACHINES CORPORATION (IBM). In recent years *Fortune* Magazine has listed IBM as one of the top ten largest corporations in the United States. It is acknowledged to be the largest and the single most influential organization within the data processing industry. It is also one of the best run companies in the world. Its story provides many insights into the development of data processing in general. In fact, many historians have treated the history of data processing as a mirror of IBM's own evolution. Although such an approach unfairly downplays the activities of others, its function within the industry is nonetheless significant.

The history of IBM can be divided into three stages: (1) 1911 to late 1940s, when IBM's role in the office marketplace consisted in selling card punch equipment; (2) late 1940s to 1964, when IBM first entered the computer market and introduced the System 360* computers, and (3) 1964 to the present, when IBM became one of the world's largest and most successful companies in the information processing arena.

IBM's origins go back to the nineteenth-century world of office equipment sales and personalities. In the late 1800s the world of information processing was coming alive with new technologies to handle large amounts of data. The cash register and the typewriter were becoming increasingly common tools of business while the telephone hinted of more rapid and interesting communications. But the real information-handling revolution came from the work of an engineer working in Washington D.C., Herman Hollerith,** an inventor who developed card punch equipment. This equipment permitted the U.S. Bureau of the Census to take the Census of 1890 faster and to analyze the data gathered more efficiently. As a result, other governments began using card-tabulating equipment in the late 1890s and early 1900s for census-taking and other applications that involved large amounts of information. Large companies handling great quantities of data, such as railroads, banks, and insurance firms, also began to use such equipment. Government agencies tracking population and gathering health statistics and tax information increasingly began to rely on such technology by the early 1900s. Vendors appeared, marketing keypunch machines, sorters, and tabulators. The most important of these firms for the history of IBM was Hollerith's own Tabulating Machine Company.

In addition to his firm, which ultimately became part of IBM, a number of others were merged together into one firm by Charles R. Flint,** a turn-of-the-century builder of conglomerate businesses. The International Time Recording Company of New Jersey, which made clocks for businesses in the early 1900s, ultimately became part of IBM. Flint also established the Computing Scale Company which made light scales such as a butcher might use. In 1911 Flint joined all these little companies together—probably more to take advantage of stock speculation than to build a great corporation—and called it the Computing-Tabulating-Recording Corporation, or simply C-T-R. Assets amounted to $17.5 million. Over the next several years, sales remained flat, and he made no attempt to consolidate any products or organizations. Then on May 1, 1914, the company hired a new general manager: Thomas J. Watson.**

Watson immediately began to invigorate the firm along the lines of his own experience as a general sales manager at the National Cash Register Company (NCR).* Watson had nearly twenty years of sales and management experience behind him. Smart, hard working, a visionary, and an admirer of salesmen, he quickly focused attention on the development of new products and marketing strategies, particularly for that portion of the business that had been part of Hollerith's operation and that Watson believed had the greatest future potential. He put together a sales force of the calibre he had seen at NCR and built a corporate culture that would later be seen as hallmarks of IBM, including the THINK logo, conservative business dress, dominance of sales in business decisions, continuous product development, attention to the welfare of employees, and the cultivation of customer contacts. Watson was very successful. Between 1914 and the end of 1917, sales increased from $4.2 million to $8.3 million. The 1920s loomed ahead as a period of prosperity and opportunity, as in fact it turned out to be. The demand for information-handling equipment grew, and Watson was in a good position to play a major role in that American market.

Watson disliked the name of the company he ran and sought one that would more accurately define its role and reflect the future he envisioned. In February 1924 C-T-R became the International Business Machines Corporation. It quickly acquired its more familiar title, IBM, and a corporate symbol of a globe spelling out the name of the firm. The new company prospered in the 1920s. Sales rose from over $10 million at its birth to $19.7 million in 1928. Although sales dipped in the 1930s during the Great Depression (to a low of $17.6 million in 1933), IBM entered 1940 with sales of $46.3 million.

The company prospered despite competitors, primarily the Powers Accounting Machine Corporation,† NCR, Remington Rand,† and Underwood Elliott Fisher. Watson was successful because he concentrated on the sale of information processing equipment to businesses and government agencies while turning out new and improved models of his products. Salesmen proposed systems of card punch equipment to tackle specific business problems important to their customers. IBM made sure that equipment was well maintained, while salesmen and their managers kept a running dialogue with their customers about new "applications" for these products. IBM also sold the necessary cards for use with its equipment—over 4 billion in 1935 at $1.05 per thousand. During the 1930s, the installed base of rental machines hardly dropped, although the sale of cards did. Hence, IBM lived through the Depression better than other firms. Watson continued to develop and manufacture products, even if that meant putting them into warehouses in anticipation of the day when the economy would improve. Plants and laboratory facilities at Endicott and Poughkeepsie, New York, therefore, kept functioning during the hard years of the 1930s. During the second half of the 1930s, IBM marketed a great deal of equipment to Roosevelt's New Deal government for such applications as recordkeeping at the Social Security Administration and at the National Recovery Administration (NRA). One of the main reasons why IBM won the Social Security Administration contract was precisely because it could deliver quickly large numbers of machines in order to carry out the law creating that government agency.

Thus, a combination of persistent marketing, product availability, and customers retaining their equipment on rent allowed IBM to survive through 1935 and to prosper in the second half of the decade. IBM emerged from the 1930s as the largest, most profitable manufacturer of business machines in the United States. In 1939 its profits exceeded those of its top four competitors combined. IBM was by then a well-established firm admired by the business community and capable of thriving in the fast growing office equipment market (Table 27).

Until the late 1940s, IBM's product line consisted primarily of card punches,

Table 27
IBM Revenues, 1922–1939, Selected Years (Millions of Dollars)

1922	10.7
1928	19.7
1931	20.3
1933	17.6
1934	20.9
1936	26.3
1937	31.9
1938	34.7
1939	39.5

SOURCE: *Moody's Industrial Manual*, 1930–1940; other useful tables may be found in Robert Sobel, *IBM: Colossus in Transition* (New York: Times Books, 1981), passim.

readers, sorters, tabulators, and an electric typewriter first marketed in the early 1930s. IBM continued to sell cards and during World War II made rifles in Poughkeepsie. By the end of the 1930s most of C-T-R's products were no longer in the product line, and the last of the nondata processing items would be gone by the end of World War II. Along the way the company gained considerable experience with mechanical devices and, during the 1930s and early 1940s, with electromechanical developments. Various managers within the firm had also expressed some interest in the almost nonexistent field of computers. During World War II IBM sponsored research at Harvard University on the Mark I, created by Professor Howard Aiken** working with engineers at the Endicott laboratory. Earlier, IBM had supported work on the expanded use of card-tabulating equipment at Columbia University and funded the establishment of a research lab there. Various government projects and efforts at American and British universities leading to the development of the electronic digital computer* with stored program capabilities were all that existed at nearly a dozen locations by the late 1940s. Perhaps the best known of these projects was the Moore School of Electrical Engineerings's† ENIAC* and later EDVAC* computers, ancestors of the UNIVAC I* of the early 1950s.

Despite a very narrow set of activities involving computers, following World War II there developed a growing interest in the field within IBM. Some executives, most notably Thomas Watson, Sr., were hesitant to jump into the computer business fully at the expense of the better known, still very profitable card punch market. Others, including Thomas Watson Jr.,** favored an aggressive push into the new field, a marketing arena that was difficult to gauge in the late 1940s. During this period, however, IBM did make a plunge in that direction with its first product: the Selective Sequence Electronic Calculator

(SSEC).* Announced in 1948, it was the first commercially available stored program computer and was sold to customers who wanted more functions than the calculating capabilities of the current card punch product line.

Although the SSEC was successful, it was not until the Korean War that IBM dramatically increased its commitment to computers. Despite concern about the marketability of such devices, IBM built a second one called the IBM 701 (also called the Defense Calculator) which was first installed in 1953. This was the first mass-produced computer within the company. From that experience emerged a generation of manufacturing managers who would be called on to develop procedures for building products in the 1950s and 1960s in more efficient and cost-effective ways than IBM's competitors. In the next three decades, IBM's packaging and manufacturing methods frequently allowed the company to introduce new technologies at lower costs, often giving it sizable advantages over competition.

The 701, as a general-purpose computer, created new demand for computational products for IBM and the data processing industry. Just before initial shipments of the 701 were sent out, the company decided to make yet another new machine called the IBM 650.* It was announced in 1953 and was first delivered in 1954. The 650 finally committed IBM to what was now perceived to be the rapidly growing computer field, one that would no longer be dominated by card punch technology. The company sold approximately 1,800 copies of the 650—a massive quantity considering that good sales for previous machines were measured in dozens or less. The next general-purpose computer from IBM was the 702 in the mid–1950s. Along with the 650 it gave IBM considerable experience in the field while further defining what business applications could be justified using computers.

IBM first learned about product development in electronics and then taught its customers how and why to use computers. Its early computer customers had done business with IBM for decades with tabulating machines and were now turning to computers primarily for fundamental accounting and inventory applications. Other customers were added throughout the 1950s. Customer support increased, as did the use of short-term leases as incentives to increase the willingness of companies to use computers. In particular, the free assistance given customers learning to use software was especially critical to the expansion of IBM's sales in data processing. Business volume reflected the company's success. Sales in 1952 were $333.7 million and by the end of 1956 had grown to $734.3 million.

IBM's big competitor in the early 1950s was Sperry Rand† which was selling the popular UNIVAC I. Sperry Rand had a chance to dominate the industry—indeed, the name *UNIVAC* was synonymous with the word *computer* in the early 1950s. But Sperry Rand failed to introduce follow-on products quickly enough (UNIVAC II did not come out until 1957). In addition, its lack of aggressive and total commitment to computers allowed IBM to seize the lead by 1956–1957. In 1956 IBM signed a consent decree with the government that terminated a running battle between the two over the company's previous methods of selling tabulating equipment and some computational devices.

A whole stream of products appeared in the late 1950s, indicating that the company was fully involved in the industry. By the end of 1960 over two-thirds of IBM's revenues came from the sale of computers and peripherals. FORTRAN* and COBOL* programming languages* appeared, along with the 305 and 709. IBM's STRETCH* effort led to a series of technological innovations in the late 1950s and early 1960s. The most important technological benefits included new packaging techniques for second- and third-generation components, printing of circuit boards and cards, and better wiring methods. Some of these innovations were first evident in the IBM 7070 and IBM 7090 computers, announced in 1958, as the company's initial second-generation commercial computers. The 1401 series, announced in October 1959, eventually resulted in the installation of nearly 20,000 copies. To put that achievement in perspective, in 1960 there were only 6,000 general-purpose computers installed in the United States from all vendors. IBM had learned a great deal about how to make and profitably sell second-generation equipment.

IBM had learned that to be successful in this new industry it had to introduce new products at least as rapidly as its competitors—companies working on various types of equipment and at different levels of technological innovation. Furthermore, each new machine had to be both functionally superior and less expensive than its predecessor. Much of the history of product introductions from that time to the late 1980s was essentially a story of new products meeting each of those requirements. This lesson was lost on a number of companies that tried to sell against IBM in the 1960s. IBM had not yet realized in the 1950s (but would as a result of its experiences in the early 1960s) that only companies willing to allocate all of their resources and talents to making new computers would survive. Dissipation of resources and energies spelled disaster. This was the hard lesson learned by those who failed to effectively market computer products in the 1960s. To mention two examples, Radio Corporation of America (RCA)† and General Electric (GE)† divided their attention among many areas, especially television, radio, home appliances, and computers.

But if IBM's success appeared destined to increase dramatically, to management circumstances pointed downward. By 1960 IBM was selling fifteen different product sets, almost all of which were incompatible with each other. (That is, programs on one machine could not run on another without being rewritten.) Other companies were preparing to enter the computer field with new and less expensive products, whereas IBM's were rapidly reaching the end of their marketable lives. At best, some could be expected to last to 1964 or 1965. The future appeared bleak. Faced with the threat of competitive pressures, in 1961 the company commissioned a task force called the Spread Committee to make recommendations. In December it suggested that IBM produce an entirely new family of computers, all of which would be compatible with each other in various sizes, along with a new line of peripherals that could be used with any of the new computers, and with fully developed operating systems compatible across different-sized machines. Furthermore, it was recommended that the hardware be based on what would become known as third-generation technology: transistors, new memory, and less expensive components promising new levels of price/performance and reliability, not to mention capacity. In early 1962 Thomas Watson, Jr., agreed to support the findings and conclusions of the task force. The result was the S/360,* perhaps the most dramatic success story in the history of American products, even surpassing the impact of the Ford Model T car.

The success of this product line is difficult to appreciate today because most of the common elements of any computer system of the 1980s first appeared with the S/360. Prior to the S/360, computers were not compatible, programs were written at the machine level as a normal course of events or one level up (using something like Assembler or FORTRAN), and operating systems to handle system control and data were almost nonexistent. Capacities were small and expensive. All of that changed with the S/360, and the fundamental design of that system remained the industry standard as late as the mid–1980s.

In the early 1960s in order to survive, IBM had to introduce new technologies where they barely existed before. New levels of price/performance had to be achieved. To do all of this required the full attention of the entire company. Plants were expanded and modernized, the field engineering force and field marketing had to be totally retrained, new products were developed in laboratories around the world, and top management worked nearly full time on the task. Then, on April 7, 1964, IBM unveiled the S/360—more products in one day than any computer company had announced up to that point.

IBM initially brought out five models of the S/360 along with new disk and tape drives, the 1403 N1 printer (considered the most popular impact printer in the industry for the next fifteen years), an operating system called OS, new compilers for FORTRAN and COBOL, card input/output equipment, and software migration tools to take programs running on the 1401s to the S/360s. The data processing industry instantly realized that a major breakthrough in technology and in the cost justification of data processing in general had just taken place. Business volume suggested the impact. In 1965 IBM forecast demand for 589 S/360s, but shipped 668 and had orders for 4,487. Through 1966, it had projected demand for over 6,800 computers but ended up with nearly 18,000 orders. The company rushed to expand its plants in order to meet this new level of demand. Thus, the impact of the S/360 was enormous. The industry as a whole grew by nearly 30 percent each year during the remainder of the 1960s. By 1969 over 15 percent of all computers in the world were S/360s. By the end of the first full year of shipments of S/360s (1965), IBM had captured approximately 65 percent of the computer market in the United States, with Sperry Rand with 12 percent a distant second. A new era had begun for IBM.

The nature of the company began to change. Assets, which had been four times current liabilities in 1962, had allowed IBM to finance the development of the new system. Hundreds of millions of dollars were borrowed, while current liabilities jumped from $158 million in 1961 to $1.16 billion in 1967. Revenues in 1964 (the last full year before S/360s were shipped to customers) were approximately $3.2 billion. By the end of 1972 IBM's revenues had reached $12.8 billion. It added six new plants to handle the S/360 machines. The number of IBM employees grew by a third to over 150,000. IBM had become a major manufacturing company in a very short period of time.

Across the industry, the demand for new systems and online applications instead of simply batch processing grew along with distributed processing and online availability of data stored in disk drives. S/360 led to the development of higher level languages that made programming easier and to reduced expenses for computing. Costs per transaction dropped dramatically. For example, between 1953 and 1964 IBM increased the speed of processors by forty times and memory by 6.5 times, while keeping the costs the same as with the older machines. The lease price of a 1964 computer (such as a S/360 Model 30) was essentially the same as that of a 1953 device ($650). Between 1948 and 1968 the cost of executing an instruction had dropped by a thousand to one (based on a comparison of only IBM processor lease prices in 1948 and 1968 versus the power acquired).

IBM's growth rate in the 1960s was usually about 17.6 percent. The entire industry's growth rate during the decade each year was closer to 27.1 percent. This meant that IBM's penetration of the U.S. data processing market went from 51 percent in 1961 to 34 percent in 1970 when the industry was much larger than in 1961.

The company continued to introduce a flood of new processors and peripheral equipment throughout the 1960s. By 1968 considerable competition within the industry for third-generation equipment had once again compelled the company to consider follow-on products. Fearing flat earnings otherwise, it had no choice. As it turned out, between 1968 and 1972 IBM's earnings remained essentially level while the company's major rivals generally doubled theirs (Table 28).

IBM's answer came in June 1970 when it introduced the S/370, a family of computers, software, and peripherals that continued the S/360 line with updated technology and more function at better prices. For the better part of the next decade, additional members of the S/370 were introduced. S/360s were quickly returned to IBM, at a time when the U.S. economy was in a recession, to be replaced with the less expensive S/370s. The early result was a contribution to IBM's flat earnings in the first several years of the 1970s. Yet sales improved when the economy turned upward. Major new products continued to enhance IBM's product set in the "glass house." The IBM 3330 disk drive made important technological improvements over second-generation devices as an example of the continuing change. It could transfer data 250 percent faster than the old 2314s and house nearly three times more information using far superior and more reliable components.

Important additional members of the S/370 family included the large Model 155 and 165, followed by the introduction of the Model 145—the first commercially available computer that used only monolithic semiconductor technol-

Table 28
IBM Revenues, 1941–1964, Selected Years (Millions of Dollars)

1941	62.9
1945	141.7
1946	119.4
1948	162.0
1950	214.9
1951	266.8
1953	333.7
1955	563.5
1956	734.3
1957	1000.4
1959	1309.8
1961	1694.3
1964	2306.0

SOURCE: Annual issues of *Moody's Manual of Investment*, also known as *Moody's Handbook*, and *Standard & Poor's Guide*.

ogy for its main memory. The machine was five times faster than its predecessor, the S/360 Model 40, with an improved cost for processing of nearly 300 percent.

Throughout the decade other products appeared, the more important of which were a series of terminals called the 3270s, 3705 communications controller to help manage networks, the S/370 Model 158 and 168, Virtual Memory (VM) operating system, MVS for large computers, and DOS/VS for small computers. Enhancements in technology and price/performance appeared gradually and were not clustered together as happened in 1964. This evolutionary approach was largely a reaction to competition which proved aggressive throughout the decade. The main battleground was initially the large computer market, but by 1970 it had spread to distributed processing and minicomputers.

Two characteristics of the computer marketplace of the 1970s affected IBM's position. First, there was the existence of plug-compatible equipment dealers, and, second, leasing companies. Plug-compatible equipment dealers were producing machines that were almost exactly like IBM's and that were sold to customers who had S/370 hardware at prices below the large manufacturer's. Thus, for instance, Memorex† sold 3330 look-alike disk drives, whereas Telex† copied IBM's 3420 tape drives. Another major competitor in the peripherals market was Storage Technology Corporation (STC). In 1975 the plug-compatible game became more sophisticated when S/370 compatible computers first appeared, primarily from the Amdahl Corporation.† Such machines ran IBM software and used IBM peripheral equipment and were marketed at lower prices or with more flexible terms and conditions to IBM's customers. In addition to Amdahl, others entered the plug-compatible market, including Magnuson Systems Corporation (which went after the 4300s and failed), Two Pi Company, and National Advanced Systems (NAS) supported by Japanese funding.

Leasing companies also flourished in the 1970s. First evident in the late 1960s, they had come into their own by the time IBM began shipping S/370s. Leasing companies bought IBM processors and then leased them to IBM's customers at lower rates than IBM's, often by amortizing the cost of such equipment over longer periods of time than might be done otherwise. Thus, a four-year IBM lease was replaced with five- to ten-year depreciation schedules that permitted lower monthly payments. It was big business. In 1969 alone leasing firms had acquired $2.5 billion worth of IBM's products. In the early 1970s they also sold S/360s in direct competition with S/370s and, later, second-hand S/370s against products announced in the late 1970s. Customers who signed up for a seven-year contract with a leasing company frequently found it difficult to upgrade to newer technologies and bigger or other computers, frustrating IBM's attempts to sell newer products.

Yet the effort by most leasing companies ultimately proved to be a disaster, certainly by the end of the decade. Most leasing companies had depreciated their equipment over longer periods of time than IBM had done, and, as a consequence, when the manufacturer introduced new products at lower prices, along with other firms, the inventories of leasing companies were not sufficiently depreciated to reflect true market values. Consequently, they lost money trying to preserve the value of their assets which increasingly had to be leased at lower rates than their depreciation schedules permitted for reasonable profitability. Despite this problem, these companies continued to provide IBM with several billions of dollars worth of competition during the 1970s. By the early 1980s the change in technologies made it extremely difficult for leasing companies to operate in the manner they had before. Now they needed to depreciate equipment faster and thus their rates went up.

Minicomputers also came into their own in the 1970s. A host of new vendors entered the market and prospered by selling systems that were installed outside of the traditional data centers. These included Data General, Hewlett-Packard,† Prime Computer,† Wang Laboratories, and Tandem Computers. Thus, by the mid–1970s the data processing market was far more complex than it had been in the 1960s. It was growing and so was IBM, even though the company did not expand as rapidly as its markets.

The plug-compatible situation was severe. By the mid–1970s over 70 percent of all the dollars invested in hardware was either in plug-compatible equipment or in third-party leasing arrangements. There were numerous plug-compatible hardware products for every item in IBM's sales kit. IBM responded by continuously announcing new products along with penetrating new markets (e.g., that of the mini), while holding down costs to combat second-hand computer vendors as well. Some of the more important products announced in this period included the 303X large processors (first announced in March 1977), the 8100s (announced in October 1978) for distributed processing, and then, in January 1979, the 4300 intermediate-size processors. The 4300s were a huge success—over 20,000 were ordered within the first year of introduction alone. New models of the 4300s continued to appear over the next seven years.

The 1970s were also critical years in the history of IBM because of the threat of lawsuits. IBM was plagued with antitrust litigation with many of its competitors as well as the U.S. government—so much so that one historian, Robert Sobel, tagged the entire period "a generation of litigation." It began in December 1968 when Control Data sued IBM for violating the antitrust laws, arguing that IBM had not honored its commitments under the consent decree of 1956. CDC accused IBM of trying to gain a monopoly within the industry. The legal profession now entered a new era of massive work with IBM's lawsuits. The CDC case alone involved IBM's lawyers examining over 120 million CDC documents while CDC studied between 25 and 40 million IBM documents. The joke at the time was that the only beneficiary of this lawsuit would be Xerox* which was selling copiers to the lawyers representing both firms. IBM countersued, and finally, in January 1973, the two settled out of court. Both destroyed the massive indexes they had developed while examining each other's documents.

The U.S. Justice Department posed a much greater danger to IBM, for a government victory would have led to the company's breakup into several firms. The antitrust division of the Justice Department had been examining the possibility of suing IBM since 1965. Its concern was that IBM may have made it difficult for new companies to enter the market and that the firm was attempting to monopolize the industry—charges which IBM denied. Then in January 1969, on the last working day of the Johnson administration, the government filed suit against IBM. The history of this suit is a long, complicated one concerning hundreds of lawyers working for the defense, the examination of billions of documents, expenditures of millions of dollars, and, in the end, during the first term of the Reagan administration, the conclusion of the case. Upon examination of the evidence gathered, the Justice Department concluded that it had no real case against IBM. The whole case had turned on the issue of what constituted the computer marketplace and in defining IBM's role. The government's definition was too narrow. The data processing industry had expanded in many ways during the 1960s and 1970s, with many new firms now in leadership roles, making any contention regarding a position of dominance difficult to establish.

The government may also have been discouraged by the fact that other cases were coming to an end. Many competitors sued IBM in the late 1960s and early 1970s in hopes of winning damages of massive amounts, but they were all defeated. The *Greyhound v. IBM* case was thrown out of court in 1972. Additional litigation relating to that case continued throughout the 1970s, however. The *Telex v. IBM* fight ended when the court sided with IBM. CalComp's suit was dismissed by another court in 1977, and so it went for over a dozen antitrust suits with juries or judges either unable to determine what the marketplace was like and hence how IBM or any other firm could monopolize data processing or, more frequently, to ascertain that IBM simply had not violated either its consent decree or any antitrust laws. Thus, along with its own internal examination of the case, the government concluded that further litigation was useless and, therefore, dropped it in January 1982. There was one small consolation from this very complicated case. Historians benefited from all these lawsuits because millions of documents relating to the history of data processing became public, covering the industry's activities from the 1950s to the early 1970s.

While IBM was selling the S/360 and later the S/370, and experiencing a great deal of competition, rapid growth, technological innovations, along with fighting lawsuits, its worldwide operations expanded. IBM had been selling equipment in Europe and in Latin America since the early 1900s. Hollerith, for example, had actively sold his machines from Great Britain to Russia. By 1930 there were sales offices in Europe, along with manufacturing facilities in Germany, France, and Great Britain. In the 1930s the company established offices in Asia and Latin America, and by 1949 all non-U.S. sales accounted for approximately 5 percent of the company's sales. That year IBM was divided into two parts, a U.S. firm to handle sales within the United States and another to manage sales everywhere else. IBM outside of the United States was named the IBM World Trade Corporation and was placed under the management of Dick Watson, son of Thomas J. Watson, Sr. During the following decade, more sales offices were established around the world, along with manufacturing and research facilities working in concert with U.S. locations.

The market was competitive and nationalistic. For example, in France the government encouraged the development of its own fledgling computer industry (primarily through Machines Bull) and urged customers to buy French-made products rather than from the American IBM company. IBM responded with aggressive pricing actions and continued to implement its established policy of staffing each national company with local employees. Thus, IBM Italy would be run by Italians while IBM Japan was managed by Japanese. The World Trade Corporation flourished, though in a smaller market than the U.S. market (Table 29). Between 1955, when World Trade's revenues totaled $132.8 million, or nearly 19 percent of all earnings that year, and increased to $1.085 billion in 1965, representing nearly a third of IBM's total. European sales provided two-thirds of World Trade's revenues throughout the 1960s. By the early 1970s, IBM had offices in 130 countries and was frequently cited as an example of a transnational corporation—an image denied by the company, given the national heritage of its employees in each country.

Because of the enormous growth of the World Trade Corporation in size and complexity, the foreign operations were split into two separate parts in the early 1970s. European, Middle Eastern, and African operations were joined, while the country organizations in Latin America and Canada were coupled to Asian companies to form yet another corporate structure. The European, Middle East-

Table 29
IBM World Trade Revenues, 1955–1965, Selected Years (Millions of Dollars)

1955	132.8
1957	202.1
1961	497.6
1964	933.0
1965	1085.5

SOURCE: Robert Sobel, *IBM: Colossus in Transition* (New York: Times Books, 1981): 197. Copyright © 1981. Reproduced by permission.

ern, and African portion accounted for over $3.5 billion in annual sales in the early 1970s and was under the command of a career French IBMer, Jacques Maisonrouge. In the early 1970s, when a recession hit the domestic company hard, expanding sales in Europe allowed IBM as a whole to enjoy continued growth. By the late 1970s the data processing industry had fully recovered and was expanding at over 12 percent each year in the United States.

Because the company was growing slower than the rest of the industry during a period when technological change was occurring more rapidly than in the 1960s or even most of the 1970s, IBM established new business objectives for the 1980s. It made a decision to compete in all segments of the information industry, and to be the low-cost producer and provider of quality products. The chairman of the board, John Opel,** also mandated that the company grow as fast as the industry. To accomplish each of these objectives new tactics and products were needed. In the late 1970s IBM began investing $17 billion in modernizing and expanding plants. It launched another aggressive product development program which resulted in thousands of new products being introduced in the early–to mid–1980s. To improve the quality of marketing services to its customers, in October 1981 IBM reorganized (effective January 1, 1982) its three U.S. marketing divisions into two.

IBM expanded its use of independent business units (IBUs) to experiment with new markets. IBUs were small organizations within IBM, not fettered by the bureaucracies of a division, which had their own board of directors, and experimented with new products and marketing techiques in areas of data processing new to IBM. This approach proved highly successful. Out of sixteen such experiments of the late 1970s and early 1980s, only two were closed down as projects not worth pursuing. The others either continued to expand as IBUs or were folded back into more traditional divisional structures once they had proven their worth.

The IBUs played an important role in IBM's enormous successes of the early 1980s. The most dramatic example of this process at work involved the IBM Personal Computer (PC). The market for microcomputers for home use or as intelligent terminals in an office was small in the 1970s. By the late 1970s, however, there were some vendors, such as Osborne† and Apple,† selling microcomputers and thus helping to define a new market. After understanding what the potential market could be, IBM decided to introduce its own product. It first established a task force to make recommendations. The task force, under the direction of Philip D. "Don" Estridge,** urged that IBM buy components from outside its own plants, assemble the machine in IBM's facilities, and make public the machine's architecture so that microcomputer manufacturers and software writers would be encouraged to make products compatible for this device. The second major step involved accepting those recommendations, with Estridge heading up a new IBU.

Within twelve months after the project was approved, IBM announced the Personal Computer. The product instantly defined a market segment as dramatically as the S/360 had seventeen years earlier. It brought standardization to the microcomputer marketplace, encouraged the use of such devices from all vendors both in homes and in businesses, and, by 1984, accounted for 4 percent of IBM's revenues. Approximately 22 percent of all microcomputers that year came from IBM. The name PC became synonymous with small computers. It opened a boom market involving millions of processors, billions of dollars, over 150 competitors building micros, and several thousand organizations marketing software packages. Microcomputers had increasingly become a visible part of many homes and even more offices and plant floors.

When the PC took off, a manufacturing plant for the product was built in Boca Raton, Florida, and a new division was established there called the Entry Systems Division, to handle all of IBM's microprocessors. By 1986 the PC product line alone had grown to over a dozen models and to hundreds of other related items. Thousands of software packages were available for it, and PC-related magazines were being published by the dozen. Some of the magazines were hundreds of pages long and appeared each month.

In addition to using IBUs to launch new business ventures, IBM sought other channels of distribution for its products over and above its sales force. The strategy that developed involved using salesmen to market the entire product line to the company's largest customers in an integrated, coordinated fashion on a nationwide basis while relying on dealers to move products into firms wanting to buy few quantities of low-cost items. Such outlets also served individuals buying IBM products. Thus, for example, IBM signed an agreement with both ComputerLand† and Sears to sell PCs in their retail outlets. Value-added remarketeers were also cultivated. Under that kind of an arrangement, a vendor

Table 30
IBM Revenues, 1965–1985, Selected Years (Millions of Dollars)

1965	2,487.3
1969	7,196.0
1972	9,533.0
1976	16,304.0
1979	22,863.0
1982	34,000.4
1985	50,000.0

SOURCE: IBM, *Annual Report*, passim.

would buy PCs or, for instance, the S/36 or Series 1 (two minis), add to them software for particular applications, and resell the hardware and software as a package deal to other companies. IBM introduced new terms and conditions in its contracts to allow large customers to buy products in quantity with various discounts based on volume. This last change—quantity discounts—represented a major departure in strategy from the past when the price for a product was always the same, no matter how many a customer purchased. With flexible terms and conditions, IBM was able to sell products within a rapidly growing data processing industry both more quickly and very competitively.

The result of these tactics was a revitalized IBM which in the 1980s was growing faster than it had in the late 1970s (see Table 30). It closed 1984 with worldwide sales of some $46 billion and anticipation of growth at or above 15 percent compounded each year. Industry analysts were forecasting that, if those growth rates continued by the mid–1990s, IBM would have sales of some $185 billion, thereby making it the largest corporation in the world's largest industry.

The 1980s witnessed other changes at IBM. By the end of 1984, for example, IBM employed over 392,000 people and expected its headcount to continue rising throughout the decade. IBM was growing at or ahead of the U.S. data processing industry, an economic sector which in 1984 accounted for nearly 3.5 percent of the gross national product. In February 1985 the 308X large computers were enhanced with the announcement of the 3090 Models 200 and 400, while the 4361 and 4381 had become the stars of the intermediate marketplace. Its S/36 mini was broadened as a product set throughout 1984 and 1985, with many customers buying them hundreds at a time.

In line with IBM's strategy of marketing in every field of data processing, it turned its attention quickly and aggressively in the direction of telecommunications in the early 1980s. Technology was rapidly causing data processing and voice communications (such as telephone technology) to merge into one. PBXs for switching telephone calls were becoming more sophisticated computers that expanded to handle text in addition to its traditional voice message. Computers in turn were moving in the direction of handling voice. IBM decided to become a major force in what in the early 1980s was considered to be a market equal to or greater in size than data processing but one that also would be a blur of the two. In order to gain quick expertise and a solid foothold in the telecommunications arena, IBM acquired the Rohlm Corporation fully in 1984, consolidated all of its internal telecommunications projects and related IBUs together, and expanded its product and marketing efforts to join data and voice. Its major competitors now began to be Northern Telecom, Ltd.† and American Telephone and Telegraph (AT&T).†

No history of the data processing industry could be written solely through the experiences of one firm. Yet IBM's story highlights many of the characteristics of that economic sector if for no other reason than because of the company's size and technological achievements over the past thirty-five years. The introduction of many computers in the late 1950s, standardization of computer systems brought about by the S/360 and later the PC for microcomputers, use of operating systems along with network and database managers developed by IBM's software community—all brought change and direction to a new industry. These products and the way they were sold and used defined data processing for well over a third of the twentieth century. Students of American business have recognized that fact. More books and articles have been published on IBM than on any other corporation in the world. Most histories of data processing either devote large sections to a discussion of IBM or are completely devoted to the company.

IBM's success can be summarized quickly with several brief observations. First, it has always been a well-managed company that concentrates on shrewd selling and on being very close to its customers. Second, when IBM committed itself to tabulating office equipment and later to computers, it gambled all of its resources and energies to insure that its involvement was total. In short, the concept of "bet your company" was an all too familiar one within IBM. Third, it was willing to change, to introduce new technologies, and to reorganize when the business case was understood. The remarkable story of IBM, like that of data processing in general, did not occur by accident but by the confluence of technological innovation, price/performance improvements, and good marketing.

For further information, see: Charles J. Bashe et al., *IBM's Early Computers* (Cambridge, Mass.: MIT Press, 1986); T. G. Belden and M. R. Belden, *The Lengthening Shadow: The Life of Thomas J. Watson* (Boston: Little, Brown & Co., 1962); S. Engelbourg, *International Business Machines: A Business History* (New York: Arno, 1976); Franklin M. Fisher et al., *Folded, Spindled, and Mutilated: Economic Analysis and U.S. v. IBM* (Cambridge, Mass.: MIT Press, 1983) and his second study, *IBM and the U.S. Data Processing Industry: An Economic History* (New York: Praeger Publishers, 1983); Nancy Foy, *The Sun Never Sets on IBM. The Culture and Folklore of IBM World Trade* (New York: William Morrow & Co., 1975); *IBM Journal of Research and Development* 25, no. 5 (September 1981) which devoted the entire issue to technological histories of IBM's products; D. W. Kean, *IBM San Jose: A Quarter Century of Innovation* (San Jose, Calif.: IBM Corporation, 1977); R. Malik, *And Tomorrow the World: Inside IBM* (London: Millington, 1975); Thomas J. Peters and Robert W. Waterman, Jr., *In Search of Excellence: Lessons from America's Best-run Companies* (New York: Harper & Row, 1982); Francis G. "Buck" Rodgers and Robert L. Shook, *The IBM Way* (New York: Harper & Row, 1986); William Rodgers, *Think: A Biography of the Watsons and IBM* (New York: Stein & Day, 1969); Robert Sobel, *IBM: Colossus in Transition* (New York: Times Books, 1981).

INTERNATIONAL DATA CORPORATION (IDC). IDC is the largest consulting and publishing organization within the data processing industry. It is also publisher of the industry's most widely read newspaper, *Computerworld*, and of many of its leading popular journals. It was the first important consulting firm to conduct surveys and to do research for clients in the field of data processing. In 1984 its annual sales exceeded $240 million.

The company was founded by Patrick McGovern in 1964. McGovern, a graduate of MIT in 1959, had tinkered with computers prior to attending college and had served as editor of his high school newspaper. While at MIT he had a part-time job with *Computers and Automation*, and following graduation, he went there to work full time. In the next three years he expanded subscriptions from 5,000 to 16,000. In 1964, recognizing a need for consulting and publishing services within the data processing industry, he formed his own firm. The result was the International Data Corporation, which one reporter, Maura McEnaney, calls "the granddaddy of computer market research firms." IDC was the original company but later became a subsidiary of his newer holding company, IDG. IDC was first established in Boston with two employees in 1964 and by the end of 1984 had generated annual sales of $25 million, working out of seven locations in the United States and thirteen in other countries. In 1984 IDC had 300 employees and within IDG as a whole nearly 1,400 worldwide.

The company experienced immediate success. By the early 1980s it claimed approximately 25 percent of the consultant's market within the data processing community. IDC had estimated the size of consulting within this industry as approximately $150 million. It quickly expanded its services from its original purpose of surveying the industry with the use of high school students in the early 1960s to a broader series of services throughout the late 1960s into the current decade. By the early 1980s these included surveying data processing, communications, office automation, and software while monitoring trends within the industry, ranging from developments in technology to the strategies of such vendors as International Business Machines Corporation (IBM)† and Apple Computer† and to what was happening in the management and use of data processing.

In 1985 this large organization had 3,000 clients, about 60 percent of whom were vendors trying to understand market shares and trends. Another 30 percent were users of data processing technology from the *Fortune* 500 list. The remaining 10 percent came primarily from the financial community. To help its clients, IDG offered such services as telephone consulting, some 250 reports published in 1984 alone on specialized topics, more narrow consulting, and a large variety of publications. It has been a successful venture, frequently reaching a growth rate of nearly 30 percent a year. By 1985 a 30-percent growth had become a corporate objective. Even in the early years, sales had grown rapidly. For example, by the end of its third year, IDC had annual sales of $500,000.

IDG was perhaps best known because of its publication of the data processing industry's largest and most important newspaper—*Computerworld*. It was launched in 1967 and by 1984 was the single largest trade publication in any industry in the world with a paid subscription of nearly 120,000. Its advertising revenues were approximately $42 million in that same year. This weekly newspaper carried information on all aspects of the industry ranging from reviews of products to news about vendors, users, applications, industry issues, people, economics, government policies regarding data processing—in short, anything related to computers and data processing. This publication was in tabloid format with nearly 100 pages per week. Other publications from IDG included *InfoWorld, PC World*, and *80 Micro* (fifteen alone in the United States). A total of forty publications in seventeen other countries also existed by the early 1980s.

By the mid–1980s IDG consisted of two clearly defined components: (1) IDC which was the original organization, doing research for clients on data processing; and (2) publishing concerns which, in 1983, logged in revenues of $85 million worldwide. Its publications were managed by a subsidiary called CW Communications, and research came under the management of another company called Interdata. The company remained privately held with McGovern owning 85 percent of IDG in 1983 and other shares owned mostly by employees.

For further information, see: Robert Levering et al., *The Computer Entrepreneurs* (New York: New American Library, 1984).

INTERNATIONAL FEDERATION FOR INFORMATION PROCESSING (IFIP). This association of societies represents various communities within the data processing industry from around the world. Since its establishment in 1959, it has grown to thirty-nine national societies. Its purpose is to promote information science and technology and international cooperation within data processing. It stimulates research and development and the dissemination of information about data processing through its various conferences.

In June 1959 UNESCO sponsored the First International Conference on Information Processing in Paris, a meeting promoted by its chairman, Howard H. Aiken,** and by Isaac L. Auerbach,** both of the United States. Many industrialized nations in the late 1950s were becoming increasingly aware of a new industry called data processing, as reflected in the establishment of many societies within the industry between the late 1950s and early 1960s, most of which have remained in existence until the present. The desire to share information had been reflected in numerous academic conferences held since the 1940s, but by the mid–1950s there was also a need to share information relevant to users of computers, not simply among their inventors. Hence, demand grew for international conferences. That of 1959 simply fed that need, and it was proposed to UNESCO which was an appropriate sponsor. As a result of this meeting, other international conferences on data processing were planned. Bylaws for an organization were drafted and adopted, bringing IFIP into existence on January 1, 1960. In the early years Auerbach served as its first president (1960–1965), followed by a succession of leading figures from the international data processing community.

Over the years IFIP has established international committees to concentrate on topics and projects ranging from programming to databases, operating systems to intelligent systems, to mention a few. In 1967 the organization established a subgroup called the IFIP International Applications Group (IAG) to satisfy the needs of those concerned with administrative data processing. Its primary focus is on the use of data processing in public and business administration. A second subgroup was formed in 1979 called the International Medical Informatics Association (IMIA) to look at the needs of data processing in medicine. IFIP continued to remain affiliated with UNESCO and established connections with other international bodies related to economic development, medicine, and the United Nations in general.

The IFIP holds international conferences each year in different cities where the papers presented are primarily on data processing technologies and about their use. In addition, these conventions serve as opportunities for the exchange of information about the activities of member societies.

For further information, see: Isaac L. Auerbach, "International Federation for Information Processing," in Anthony Ralston and Edwin D. Reilly, Jr., eds., *Encyclopedia of Computer Science and Engineering* (New York: Van Nostrand Reinhold, 1983): 797–799 and his "The State of IFIP—Personal Recollections," *Annals of the History of Computing* 8, no. 2 (April 1986): 180–192.

ISA (Instrument Society of America). This nonprofit association was founded in 1945 to provide education in the field of instrumentation and control. From almost its beginning, ISA commented on the use of data processing technology in advancing the field of instrumentation, particularly in industry. It has held a wide range of seminars, conventions, and classes, and has published books, pamphlets, journals, and videos on technical subjects. ISA's membership in 1985 approached 35,000, made up of engineers and design specialists (30.6 percent), general business management (13.6 percent), sales (14.2 percent), and others from operations, maintenance, procurement, and education.

ISA is headquartered at Research Triangle Park, North Carolina. Beginning in the mid–1950s, this site became an important technical corridor. By the 1970s it was home for one of International Business Machines Corporation's (IBM's)† largest manufacturing plants, three universities, and numerous high-tech-based companies, all employing large numbers of engineers and data processing professionals. Local chapters all over the United States, called sections, held meetings and in the mid–1980s comprised nearly 200 separate entities. These in turn were part of a regional organization called a district, of which there were thirteen by 1985.

Over the years ISA has expanded its areas of interest for which it provides educational materials; there were twenty-four by the mid–1980s. Some of these divisions cover such topics as mining and metallurgy, pulp and paper, water and wastewater, analysis instrumentation, automatic control systems, computer technology, cryogenic instrumentation, process measurement and control, and scientific instrumentation and research. Essentially clustered into two large groups of subjects—the industries and sciences department and the technology department—all courses, seminars, and publications concern the twenty-four divisions. Each focuses heavily on the role of computers and other data processing technology as it applies to a particular area. The primary attention is always on improving the skills of engineers and other technicians in these fields and in establishing formal standards of professionalism for each. By 1983 ISA had a budget of over $7 million and an income of $8.8 million. In 1984 income had grown to $10.6 million, suggesting that this association was still growing and meeting a need within the technical communities of the industrialized world.

For further information, see: Instrument Society of America, *Annual Progress Report, 1983–1984* (Research Triangle Park, N.C.: ISA, 1985) and its *Training: Programs, 1984–1985* (Research Triangle Park, N.C.: ISA, 1985).

ITEL. This company was one of the most important computer leasing firms in the United States from the late 1960s through the 1970s. It acquired computers from such companies as International Business Machines Corporation (IBM)* and then leased them to users at a profit. This was a multibillion dollar segment of the data processing industry, based overwhelmingly on IBM's computers. Itel was spectacular in its early successes and equally dramatic in its departure from the industry, trapped by faulty calculations about the rate of new technological introductions and consequently improperly accounting for its depreciation of purchased inventory of computers.

Most lease-back arrangements in the 1960s concerned IBM's S/360* computers. The system was widely accepted by customers, for each of the models was relatively compatible up and down the product line, as were the peripherals and operating systems that constituted the rest of this family of computers. The machines were therefore attractive financial investments. Because IBM had made enormous investments in new technology, had rebuilt plants, and had hired thousands of workers, leasing companies assumed that the firm would keep this product line current for many years without replacing these machines until full profits could be earned and the cost of investments recovered. Leasing companies therefore offered longer term leases than they might otherwise. They also offered contracts that were longer than IBM's. The 1956 consent decree signed between IBM and the government forbade leases of over a year, meaning that if someone wanted to lease a machine for longer than twelve months, he or she would either have to buy the device outright or go to a leasing company for a longer term contract.

In the 1970s IBM was able to offer longer term leases (four years in length), but leasing companies had more flexibility to make them yet longer (sometimes ten-year deals were signed) and hence reduce a customer's monthly cash flow (payment) for the equipment. The U.S. Investment Tax Credit (ITC)—made available in order to stimulate capital investments in modern equipment—was a credit against taxes up to 7 percent of a new piece of equipment's purchase price in the 1960s and was used by leasing companies. In the 1970s this ITC amount went as high as 10 percent, making it even more attractive as a financial tool. ITC made lease-back deals attractive as a way of making money on machines while reducing a customer's monthly lease price on a computer because the leasing company would "take" the ITC (claim it as a tax deduction—credit—on its returns) and lower monthly payments on a capital lease by a proportion of that benefit generated by that tax credit. Hence, both parties found ITC a useful device.

Leasing companies bought computers from IBM and depreciated them over a longer period of time than did the manufacturer (up to ten years versus IBM's practice of seven, five, or fewer years). The differences in cash flows led to enormous profits in which Itel participated, making the stocks of such companies very attractive from the late 1960s to the late 1970s. Itel and others were therefore able to raise billions of dollars in capital to fund additional acquisitions. Some of the more important leasing companies of the time included Greyhound, Boothe Computer Corporation, and Itel. The history of Itel is a roller coaster story of fantastic successes and rapid failures which many leasing companies experienced.

Itel was founded in December 1967 when it was incorporated as SSI Computer Corporation but was simply called Itel. In March 1968 it signed the first lease and by the end of the year had leases on equipment purchased at $130 million. Itel owned nearly $104 million of this amount. All of this work was started with one salesman; the marketing force grew to no more than eight by Christmas 1968. By the end of that year Itel had become the ninth largest leasing company possessing IBM's computers. This portfolio of S/360s grew to $195.5 million in the next twelve months. With so much inventory tied up in leases, Itel's revenues grew at a steep rate from $9.6 million during its first year to $38.7 million in 1969. These U.S. revenues grew to $46.9 million in 1970—a down year for the U.S. economy as a whole and for the data processing industry in general.

During this period Itel offered leases of between two and five years, with most being closer to five. By then IBM was offering two-year leases. If a customer was willing to commit itself to a computer for a longer period of time and did not want to purchase a computer, Itel's leases were attractive as a "gap" filler. Itel would also enable a customer who had installed a computer on lease or rent from IBM (rent was on a month-to-month basis with a ninety-day notice of intent to terminate a contract by the customer) could buy the machine from IBM, cashing in purchase accruals. That customer could then turn around and sell the equipment to Itel, lease it back for lower monthly rates—all in one day and without taking out the computer.

Itel did business in every segment of the American economy: utilities, transportation, insurance, oil, manufacturing, chemicals, publishing, aerospace, banking and finance, medical services, and the food industry. It was a highly respected vendor of leases by 1970. Its successes in such a short period of time encouraged management to diversify within data processing. In its second year of operation it bought Statistics for Management Data Processing Corporation, which was a service bureau. It established Diablo Systems, Inc., to make peripheral equipment to work with its leased computers from IBM. Diablo's first project was to design and build mass memory hardware. In 1970 Itel bought Intercontinental Systems, Inc., which manufactured word processing equipment, terminals, and various other devices. Then in 1971 it acquired Information Storage Systems, its most significant acquisition which, in July 1973, it sold to Sperry Rand.† Diablo was sold off to Xerox† in 1972 for a profitable sum.

Like other leasing companies, Itel had become a major competitor to IBM in a very short period of time. Then in 1970 IBM announced the S/370* family of computers, thereby quickly making the S/360 obsolete to the surprise of numerous leasing companies. They responded, as did Itel, by lowering prices on their inventory of S/360s and by selling off other businesses in hopes of making such equipment price competitive with the new machines when they began to be installed in the early 1970s. Leasing companies started to put together package deals involving non-IBM peripheral equipment and non-IBM software (such as EDOS, a competitor to IBM's DOS software). These tactics worked well and insured that Itel and others would be able to compete with their backlog of S/360s. In the meanwhile, they also began to acquire the newer machines. During the 1970s Itel gradually discarded S/360s while shifting increasingly to an inventory of the S/370s.

But all these tactics simply covered up long-term problems that existed across the entire leasing business in the 1970s and that, by the end of the decade, were causing huge losses, certainly by the end of 1979. These leasing companies assumed they could competitively lease a product for a longer period of time than in reality was the case. They assumed "useful lives" (a term used to determine depreciation) that were too long, such as seven to ten years, while manufacturers of equipment more frequently relied on five or less. Depreciation tactics hurt these leasing companies. They would depreciate on a straight-line basis (an even amount per year over a long period of time), whereas manufacturing companies used not only shorter periods but also double-declining techniques for depreciation (the most value was taken in the first year, the second highest percent of depreciation in the second year, the third highest in year three, etc.). Therefore, IBM and others could introduce new products sooner than assumed by the leasing firms. This circumstance left the leasing companies with huge amounts of undepreciated equipment on the books that were now dropping in value, often to amounts below what remained to be depreciated. The IBMs of the industry had taken their profits and could sell newer, less expensive, better equipment. Leasing companies in turn were now forced to lower their prices too rapidly in an attempt to remain in business and competitive. In short, leasing companies had failed to recognize that their greatest revenue streams and hence profits could only come at the start of a lease and declined as time passed. Residual values (the value of equipment on the second-hand market) dropped faster than the leasing companies thought possible, so that the delta between what one could sell a machine for and what it was on the accounting books ran into the millions of dollars.

Then in early 1979, IBM announced the 4300 series of computers and was well underway with a new family of large processors (first the 3031s, 3032s, and 3033s, and later 308Xs) which wreaked havoc on depreciation schedules for their S/370s. Itel reacted to the announcement of the 4300s by lowering prices again. It also introduced a plug-compatible machine called the AS/3 Model 5, which was touted to operate 22 percent faster than a 4341. The AS/3 was part of a series marketed by Itel as a member of the Advanced Series (AS) which were built by National Semiconductor for Itel.

But these gestures did not solve the fundamental problem of depreciation, a concern that received too little attention from the industry press or within the financial community during the 1970s. Customers ignored the problem, while vendors like IBM took advantage of the cancer eating at the leasing companies. But like the last flowering before fall and winter, Itel hardly appreciated the imminence of its demise because in January 1979, within thirty days of IBM's announcement of the 4300 (it came in February), it flew 1,300 employees to Acapulco to celebrate sales in 1978, which had generated $50 million in profits. That adventure cost $3 million and was the last. At the time Itel had some of the best paid salesmen in the data processing industry, with many making over $100,000 a year, and corporate offices displaying expensive Oriental carpets.

By the end of June 1979, with thousands of 4300s on order at IBM and delivery schedules now announced, Itel now had technologically obsolete S/370s. The company posted a $50 million loss in July for the first half. By the end of the year, Itel was nearly dead. It had fired most of its top management and closed offices. Almost all of its computer businesses had been sold off or closed, and of its 7,000 employees, 6,000 were no longer on the payroll. Wall Street's excitement over Itel's stock a decade earlier turned to instant rejection, causing the company's stock to drop from $39 to $5 per share. At the end of 1979 Itel's losses amounted to $443 million. Writeoffs jumped to over $300 million. Earnings and losses were restated for the years 1978 and 1979 and for the first nine months of 1980. In 1981 the company went into receivership. Itel was no longer a factor in the industry. Lloyds of London, which had insured residual values for so many leasing companies, including Itel, stood to lose hundreds of millions of dollars as a consequence. In receivership, revenues on remaining contracts continued to flow and as late as 1983 amounted to $737 million, suggesting the magnitude of the disaster. Itel has the distinction of losing more money than any other leasing company to date in the history of the data processing industry.

Changes in tax law in the United States, lessons learned from the 1970s and the experiences of such leasing companies as Itel, and finally new rounds of technological innovations that dropped the cost of computing in the 1980s, made leasing companies less of a factor in the data processing industry than before. Their leases did not vary as widely from those of manufacturers as they had in the 1960s and 1970s. Furthermore, such data processing companies as IBM and Northern Telecom (NTL)† also set up their own subsidiaries to buy equipment from the parent company and then place them on capital leases to customers in competition with traditional leasing companies. Itel's experiences had contributed to the changes in the industry evident in the early to mid–1980s.

For further information, see: Franklin M. Fisher, et al., *IBM and the U.S. Data Processing Industry: An Economic History* (New York: Praeger Publishers, 1983); Jack B. Rochester and John Gantz, *The Naked Computer* (New York: William Morrow & Co., 1983).

J

JOINT USERS GROUP (JUG). Informally established in the late 1950s as an organization of other digital computer* user clubs, JUG was formally established in 1961 when it was recognized by the Association for Computing Machinery (ACM),† a major association within the data processing industry. Like other computer user organizations, its purpose was to share experiences and information concerning the use of digital computers.

JUG inventoried software in 1971 and again in 1974, publishing the *Computer Programs Directory*. This work listed software available in the industry and presented annotations concerning use, sale, systems, and documentation. JUG also played an active role in establishing industrywide standards for the development of software packages. Other important groups such as SHARE† and GUIDE† at one time or another became associated with JUG. More recently established user groups, such as those with products from Texas Instruments† (TI-MIX) and Wang Laboratories (SWAP), were also members of JUG.

JUG was plagued by its lack of a tight organization. (All the work was done by volunteers who were also members of other organizations.) The availability of other associations already expressing the interests and concerns of members and JUG's failure to articulate a clear direction finally led to its demise. It held its last meeting in March 1975. At that time member organizations of JUG were DECUS,† Federation of NCR† (National Cash Register Company) Users, FOCUS (Control Data Corporation† users), HUG (Honeywell users), GUIDE, SWAP, TI-MIX, USE,† and VIM.†

For further information, see: Joint Users Group, *Computer Programs Directory* (New York: Macmillan Publishing Co., 1971 and 1974); R. H. VanDenberg, Jr., "Joint Users Group (JUG)," in Anthony Ralston and Edwin D. Reilly, Jr., eds., *Encyclopedia of Computer Science and Engineering* (New York: Van Nostrand Reinhold, 1983): 811–812.

JUG. See JOINT USERS GROUP

K

KAYPRO CORPORATION. Kaypro was one of the first companies to successfully market a portable computer. Its most important products were the Kaypro II and a series of other Kaypro portable micros, including the Kaypro 10 which had a hard disk.

The company was established by Andrew Kay in 1953. Born in 1919, he was one of the oldest entrepreneurs in the data processing industry. Kay was raised in Clifton, New Jersey, and, like many of data processing's executives, he studied electrical engineering at MIT, completing work for his B.S. in 1940. After various engineering jobs, he decided to strike out on his own and build a digital voltmeter. This device tested electronic parts and whole systems of components. His company, called Non-Linear Systems, constructed a variety of voltmeters during the 1950s and 1960s. By the end of the 1960s, Kay's products incorporated minicomputers as well. During the 1970s sales ranged between $1.67 million and $6.7 million each year. As the 1970s ended, Kay found his company in a minority of less than 10 percent of the firms in the electronics industry not actively pursuing computer-based products. Furthermore, his largest and most important customers were in the aerospace industry which was then shrinking.

As a result, in 1980 Kay decided to begin marketing personal computers. Such devices had already made their appearance in California (where his company was located) as early as 1976 and various firms, including Apple Computer† and Osborne Computer,† were prospering. In 1982 he introduced the Kaycomp II which he aimed at Osborne's market. Like Osborne's machine, it sold for $1,795, and other than for a larger screen on its terminal, it appeared to be similar to it. The name of the company was changed to Kaypro Corporation, and business went so well that in 1984 Kaypro was the fourth largest manufacturer of personal computers in the industry. Its initial market was made up of engineers, but it quickly became populated by small businesses and writers. A smaller version of its micro which sold for $500 and used one of two word processing software packages also sold by the company helped. Kaypro competed with International Business Machines Corporation (IBM),† Apple, and the Tandy/Radio Shack† retail operation for that market made up of buyers of personal computers priced in the range of $1,000 to $3,000. In 1984 sales approached $175 million, although some industry watchers thought it was closer to $150 million.

The company made its first stock offering in 1983. The stock was worth $251 million for 37.6 million shares. By mid–1984 its value declined from $10 (at the time of initial offering) to just under $3 per share.

Kaypro's success was due to several factors. First, it offered a complete configuration: hardware and software. That "one price for all" approach made selling and buying easier. It was done at a time when few vendors offered both. Its prices were some of the lowest in the industry, accomplished by manufacturing components and assembling systems in one plant at Solana Beach, California. It sold products directly to dealers, thereby passing on the savings of not dealing with wholesalers. All sales to dealers were COD, which reduced the cost of accounts receivable and bad debts. These measures made Kaypro's micros relatively inexpensive.

By industry standards, Kaypro was a well-managed company. Its team spirit and participatory (consensus) management which Kay demanded were not common in small entrepreneurial operations. Rapid introduction of new products also insured the currency of the product line. Thus, Kaycomp II became Kaypro II, Kaypros 4 then 10 appeared, and finally a lap portable emerged that could run IBM PC software. One other product, called Robie, was a desktop microcomputer. By the mid–1980s the company employed 572 people. Sales rose from $75 million in 1983 to over $150 million in 1984, and from profits of $12.9 million in 1983 to nearly $23 million the following year. In 1984 considerable turnover in management and sales, along with increasing inventories, made industry analysts question the company's ability to expand even further. Yet the company continued to be important in the microcomputer market in 1985 and 1986 despite IBM's spectacular success with its own PC.

For further information, see: Paul Freiberger and Michael Swaine, *Fire in the Valley* (Berkeley, Calif.: Osborne/McGraw-Hill, 1984); Robert Levering et al., *The Computer Entrepreneurs* (New York: New American Library, 1984).

L

LINCOLN LABORATORY. This research facility at MIT produced a number of major data processing technologies in the 1950s, including WHIRLWIND,* the largest computer of the early 1950s, and SAGE,* another large project that provided the United States with an air-defense system based on computerized technology. The laboratory followed a generation-long tradition of research on computing done within the electrical engineering community at MIT. It kept MIT within the mainstream of computing projects during the 1950s and still an important element in the early 1960s.

Lincoln Laboratory was created in August 1951 with F. Wheeler Loomis as director. It was made up of divisions, each of which consisted of groups. From the beginning, it was heavily involved in research projects under contract to the military. (Division III with communications and components and Group 32 for continuous-wave components and tubes were examples). Each division of the laboratory contributed new technologies that were later used in data processing. Division VI was called the Digital Computer Laboratory and was headed by Jay W. Forrester,** father of the WHIRLWIND computer. Robert Everett,** another of the engineers on WHIRLWIND, served as associate director. Division VI was made up of six groups. Near the end of 1952 Albert Hill became head of Lincoln Laboratory. During his tenure, important research was done on behalf of the U.S. Air Force on air defense, called Project Lincoln; this work soon tied into WHIRLWIND.

Lincoln also worked with International Business Machines Corporation (IBM)† to build the AN/FSQ–7, one of the most widely used computers by the Air Force during the 1950s. Other defense-related work was conducted in concert with Bell Laboratories,† the Western Electric Company (part of American Telephone and Telegraph [AT&T]†), General Electric (GE)†, Hughes Aircraft, and Raytheon,† to mention a few. Under its wing, the Lincoln Laboratory created the SAGE system, the initial piece of which became operational in June 1958.

All the early programming for this system came out of Lincoln Laboratory. It also housed research on the TX–0 and the TX–2 military computers. Later, in the 1960s, researchers at the laboratory turned their attention increasingly to medical computing and designed the Laboratory Instrument Computer (LINK).

From the beginning then, the Lincoln Laboratory was home for much research on data processing. Richard B. Adler, for example, who headed the laboratory's solid-state and transistor group (1951–1953), had been interested in computing for many years at MIT. At this laboratory he continued research while training students in the field of transistors and their applications. His work and that of others made this laboratory one of the leading applied military research centers in the United States during the early 1950s.

One student at the laboratory, Wilbur B. Davenport, designed the SR-NOMAC computer for military applications in the 1950s. Paul E. Green, Jr., extended work on the project, building the computer. His machine implemented progress made in the general field of communications, leading to a machine called the F9C. It was constructed for the U.S. Army Signal Corps which began using it in 1953. Additional work on communications technology led to the F9C-A, also known as the RAKE.

The laboratory's most publicized role in the early years of computing related to work done on WHIRLWIND, one of the most important general-purpose digital computers* built at that time. WHIRLWIND was a major reason for establishing this laboratory. At the request of the U.S. Air Force, MIT reorganized the departments working on this computer, resulting in the creation of the Lincoln Laboratory. Project Whirlwind, as it was originally called, began in the Servomechanism Laboratory in the late 1940s, then shifted to the new Digital Computer Laboratory (under Forrester's control), and finally merged with Project Lincoln, resulting in the reorganization that gave birth to Lincoln Lab.

Of particular interest to historians of data processing was Division VI which managed WHIRLWIND and other computational projects. As noted earlier, it was divided into six groups: Group 60, which handled the laboratory's administrative services; Group 61, which managed another project called the Cape Cod Model Air Defense System which, in 1953, became a live operation for the defense of the United States; Group 62, which had the task of building the air defense computer while Group 65 worked on magnetic materials necessary for the system; Group 64, which did the maintenance on WHIRLWIND and improved on the system's design; and Group 63, which did all the work on electrostatic storage tubes. Subsequently, another organization was established within Lincoln Laboratory, called the Production Coordination Office, to work with the military on SAGE. Yet another department, formally known as the Sage System Production Coordination Office, along with the AN/FSQ–7 Systems Office, operated within the same division as WHIRLWIND.

Forrester had relative freedom to work on WHIRLWIND. He was subject more to the politics and concerns of the military than to the administrative functions within the laboratory. This was especially true in the laboratory's first two years and thus may have contributed to his ability to develop the technologies that went into the computer. Debate has arisen over the administrative environment under which the military and Forrester operated in the development of this new computer. However, it should be noted that Forrester left the laboratory to teach at MIT in 1956 and was succeeded by Everett. In 1959 Everett left MIT with some members of the division's staff to establish MITRE Corporation—a project supported by the U.S. Air Force as a further reorganization of military research into a better coordinated structure. While information on MITRE largely remains classified, it is known that it conducted considerable research on defense-related systems in the 1960s and 1970s, as well as experiments on artificial intelligence.* Division VI remained more autonomous than the other five divisions of the laboratory.

Lincoln Laboratory represented a new genre of research organizations in the United States following World War II. It became the classic example of military research conducted at an American university and perhaps the biggest of the early illustrations of such work, at least for the 1950s. Research performed there enhanced data processing's technologies, especially from 1950 to 1955.

For further information, see: Kent C. Redmond and Thomas M. Smith, *Project WHIRLWIND: The History of a Pioneer Computer* (Bedford, Mass.: Digital Press, 1980); Karl L. Wildes and Nilo A. Lindgren, *A Century of Electrical Engineering and Computer Science at MIT, 1882–1982* (Cambridge, Mass.: MIT Press, 1985).

LOTUS DEVELOPMENT CORPORATION. This company became the world's largest supplier of independent software by 1985 and a leading vendor of programs for microcomputers. It became famous for marketing a spreadsheet package called Lotus 1–2–3, the single most widely used software tool for financial and accounting analysis on microcomputers during the first half of the 1980s. Like many other data processing vendors of this period, it rose from a single product to a multimillion dollar corporation in a few years. By the mid–1980s Lotus had become a major vendor and the subject of much attention in the press.

Lotus was the creation of Mitch Kapor (1950–). In 1973, unemployed and living in Boston, he took a job as an application programmer with the task of taking data from questionnaires in a market research project and converting them into reports. In the process he learned to program in BASIC.* He subsequently flirted with graduate studies in business and continued his interest in computers but using microprocessors. His interest in these machines flowered in 1978 when he purchased first a Radio Shack TRS–80 Model I and later an Apple II. At this time he earned money writing programs in the Boston area. His first software package was developed with Eric Rosenfeld of MIT, called Tiny Troll which, among other things, prepared charts, analyzed statistics, generated multiple regressions, and edited text using a microcomputer. With the help of Robert Frankstron, a creator of VisiCalc (the first widely used spreadsheet package in the industry), Kapor signed a contract with VisiCorp to produce his own spreadsheet product similar to his first one. In 1981 Kapor's new package, VisiPlot, came out along with another called VisiTrend. Within six months he had made $500,000 in royalties, encouraging him to continue with such programming. In 1982 he sold the two packages outright to VisiCorp, picking up an additional $1.5 million.

Kapor's early successes encouraged him further, leading to his association with Jonathan Sachs, who had greater technical background than he did in programming. In August 1981, International Business Machines Corporation (IBM)† announced the PC microcomputer based on 16-bit technology. All other micros of that period were based on 8-bit chips. Kapor and Sachs concluded that 16-bit represented the wave of the future and thus decided to write a software package that would run on such technology and to do it rapidly before others had the same thought. With the help of a venture capitalist named Ben Rosen, who had personally used Tiny Troll to manage his own investments, he raised nearly $4.2 million to launch his company which was named Lotus.

Lotus was officially established in April 1982, and at the COMDEX data processing convention in the Fall, Kapor publicly announced the availability of Lotus 1–2–3. It was an instant success, selling 60,000 copies by July 1982 and over a million by early 1985. The package was an easy-to-use spreadsheet package that worked quickly. It was flexible and had the ability both to manage a database and to generate graphs. It was one of the first packages to run on an IBM PC.

Its success has also been frequently attributed to good timing. First, there was the availability of the IBM PC for which Kapor wisely wrote programs in the beginning. IBM's product became the standard among microcomputers; several million of them were sold by the end of 1985. In turn this product created a demand for Lotus 1–2–3 since the most widely used application for such machines was spreadsheet analysis. Second, the U.S. economy in 1981 experienced a resurgence that led to rapid growth in sales of all data processing products. Even after the data processing industry experienced a downturn in 1985 complete with layoffs and declining profits and sales, Lotus did well. Finally, the software turned in an outstanding performance. At the end of 1983, the first full year of the company's operation, revenues reached $53 million. In October of that year, the company went public, selling $34 million in stock. Like many other successful entrepreneurs in the industry, Kapor had become a multimillionare quickly and while still a young man.

In January 1984 Kapor's company announced Symphony, an enhancement of the original package, and made its first shipments in July. By April 1985 it had generated $40 million in sales. Symphony had modules that permitted such functions as spell checking; a module called Symphony Link which allowed communications between a microcomputer and a host mainframe; and additional work processing capabilities. The company sold other products as well. For example, Jazz was Lotus 1–2–3 for use on Apple's† portable microcomputer. Lotus also had a 1–2–3 report writer and Spotlight Desk Organizer. By the spring

Figure 5
Leading Personal Computer Software Publishers' Revenue (Worldwide)

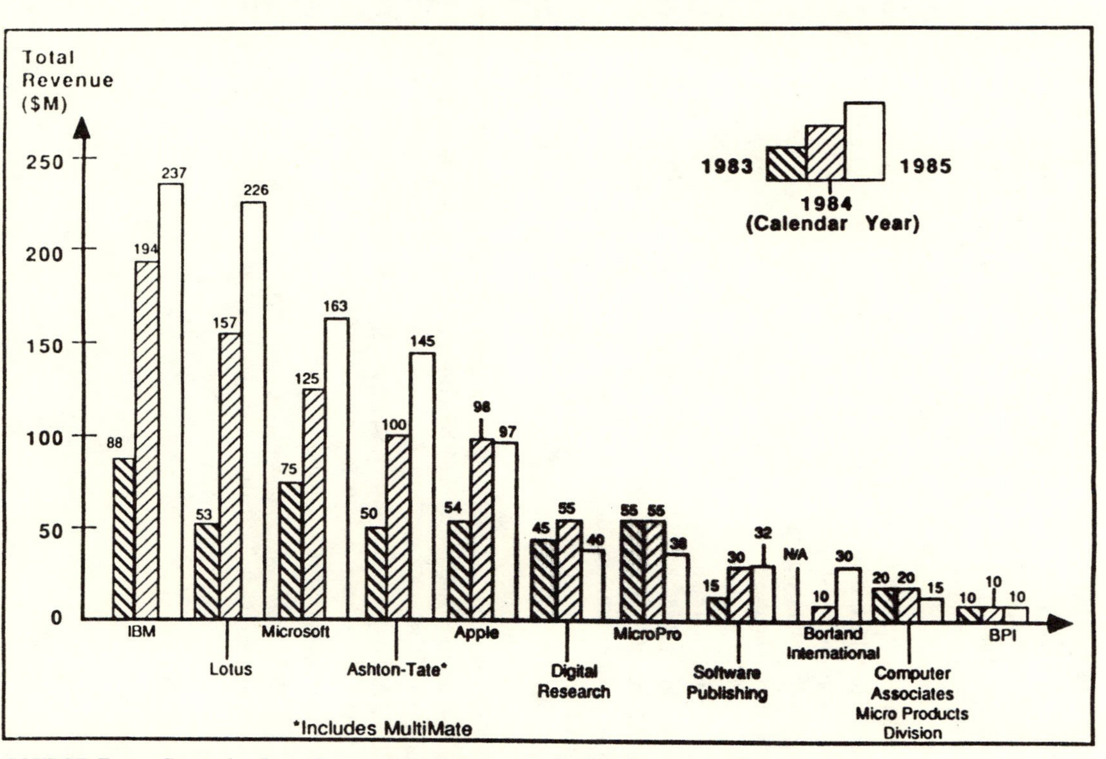

SOURCE: Future Computing Inc., *Computerworld*. Copyright © 1986. Reprinted with permission.

of 1985 all of these products were selling around the world, with sales outside of the United States accounting for 10 percent of all revenues.

Kapor used nontechnical names for his products, especially musical names, because he believed metaphors allowed individuals to make emotional attachments to data processing products rather than impersonal technical linkages. The idea for using Lotus as the name of the company and as his first major product, came from India, where it was associated with the concept of perfect enlightenment. He also had an interest in music. (During the 1970s he had taught piano lessons). Hence, his heavy use of musical names for software.

The organization of Lotus, Inc., evolved over the years according to the kind of experiences typical of many high-tech companies in the industry. Later, others began to embrace the free-form structured organization favored by Kapor and Sachs in which democracy of ideas and respect for the individual were more important than institutions. By the end of 1984, however, the company had 600 employees, and managers were given responsibility for discrete areas of the business. The unstructured, free-flowing format of the early years began to give way, and by early 1985 portions of the firm were being organized into divisions.

Managers sought to broaden their product set and services in 1985. For example, in March 1985 the Engineering and Product Division was created to make and sell software to technical users, such as engineers, who then represented nearly one-sixth of the company's customers (as measured in revenues generated). Retail outlets licensed to sell Lotus's products by early 1985 amounted to some 2,200 stores (not owned by Lotus), each with an average sales force of five people.

Lotus was the most successful of the leading independent software firms as measured in terms of revenues generated (Figure 5). The second most important in firm, Microsoft,† had significant revenues but was far behind Lotus. Yet the entire marketplace had grown substantially over the period of the early 1980s, a fact that is amazing when one considers that there was no market for microcomputer software before 1976.

For further information, see: Eric Bender, "The House That 123 Built," *Computerworld*, July 15, 1985, pp. ID/10–11, 13–15, 18–20; Robert Levering et al., *The Computer Entrepreneurs* (New York: New American Library, 1984).

M

MACHINES BULL. See BULL COMPANY

MEMOREX CORPORATION. Memorex was one of the better known plug-compatible manufacturers (PCMs) of peripheral equipment (particularly tape and disk drives) of the 1960s and 1970s, competing directly against International Business Machines Corporation (IBM).† It was established in February 1961 by four ex-employees of Ampex, a leading equipment manufacturer within the data processing industry during the 1950s. When incorporated in 1961, the four capitalized Memorex at $12,500, while Allstate Insurance Company and other investors added $1.25 million to the new company's resources. The original purpose of the company was to manufacture magnetic recording tape for both computers and commercial broadcasting equipment.

By the middle of 1965 Memorex was one of the better known manufacturers of magnetic tape. Between 1964 and 1967 it sold computer tape to Burroughs,† Honeywell,† Control Data Corporation (CDC),† Univac, the National Cash Register Company (NCR),† and Digital Equipment Corporation (DEC),†—covering a broad range of manufacturers of computer-related equipment. In 1967 the company began manufacturing disk packs for disk drives manufactured by IBM and others. Its initial disk packs appeared on the market in September 1967 and were sold through its subsidiary, Disk Pack Corporation. These were sold as original-equipment-manufactured (known as OEM in the industry) products to vendors such as IBM and CDC but also in competition against these and other firms. Memorex was able to do so because it had a large set of established customers for its tape drives who also bought disk packs. In the late 1960s its customer base totaled several thousand.

Memorex next made disk drives for other equipment dealers and sold them directly against other manufacturers. In addition to this new line of products, in 1966 the company established a subsidiary, called Peripherals Systems

Corporation, to sell key-to-disk data entry recording equipment. This subsidiary never developed the necessary hardware, but it became that part of Memorex which introduced the company's first disk drive. Called the Memorex 630, it was compatible with IBM's 2311 disk drive. Memorex sold the unit primarily to leasing companies. A company such as Management Associates, Inc. (MAI) then sold such units to its customers who used IBM S/360* computers, beginning in 1968. Memorex built similar equipment for DEC to use with its computers. This product was equipped with various interfaces so that it could be sold to other vendors for use only with their particular control units and computer systems.

In 1969 the company reorganized into two groups: one was responsible for manufacturing and marketing magnetic media, and the other, called the equipment group, managed Peripherals Systems Corporation and Image Products Corporation. In May 1969, when the reorganization went into effect, the equipment group had the 630 disk drive but no other product. Engineers in the company were then developing a microfilm system that could serve as output peripherals for computers. In the 1970s the company expanded its dealings with OEM vendors, offering more disk drives. The groups sold to OEMs because the company as a whole did not have a direct sales force. These OEM customers also paid cash. This last point proved crucial inasmuch as Memorex needed money with which to pay off debts and to expand. OEMs bought equipment outright and in turn became responsible for the expense of carrying this inventory which they generally leased to their customers. In an age when most hardware was leased, Memorex's ability to sell its products was thus of considerable advantage.

Memorex hired ex-IBM disk drive engineers along with other ex-IBMers in order to build up its plug-compatible business. Between 1968 and 1971 nearly 600 ex-IBMers were hired. The company now had a considerable body of expertise with disk drives which influenced the company's actions. Memorex brought out the 630 disk file, a unit compatible with the IBM 2314 disk drive. It was announced in 1968, and volume shipments came in the second quarter of 1969. Memorex sold 1,200 660s to Radio Corporation of America (RCA)† and obtained a large contract from Univac which was soon after canceled by the customer owing to the poor performance of the drives. The 661 appeared next as a control unit for the 660; earlier, all 660s had to use control units built by IBM, RCA, and Univac. With the 661 the company could sell an entire string of disk drives to a user of an IBM S/360 and for about 15 percent below IBM's price. The 661 was announced on December 17, 1969, and became a threat to IBM in 1970.

In 1969 Memorex also began to sell to end users directly, slowly building up a marketing force of its own. Sales of disk drives grew. In 1969, for instance, Memorex claimed that sales of disk drives reached $15 million. The number of employees increased from 1,916 in late 1969 to 6,101 by the end of 1970. In 1968 annual sales from all products generated revenues of $60 million. Sales in the United States for data processing products rose from $390,000 in 1962 to $41.5 million in 1970.

Memorex raised a considerable amount of capital to fuel its expansion in the 1960s and 1970s. Besides the initial $1.25 million, in 1962 it acquired another $608,000 in debt. By 1966 Bank of America had loaned the company $6 million. That year Memorex made a public offering worth $12 million which it used to repay debts. By the end of 1969, however, it had borrowed between $40 and $50 million from the Bank of America. Laurence Spitters, who ran Memorex during these years, used this debt to expand the company into a sizable enterprise.

Competition between Memorex and IBM heated up in the early 1970s. In 1973 Memorex filed a private antitrust suit against IBM, following the example of nearly a dozen other vendors who sought to cripple the large manufacturer. Like them too, Memorex failed to win its case. Meanwhile, Memorex expanded its IBM plug-compatible product line. In October 1971 Memorex announced the 3670 disk drive which it pitted against IBM's newly announced 3330/3830 disk subsystem. It also slashed prices on its 2314-compatible equipment. Demand for disk drives throughout the industry remained so great that both IBM and Memorex were able to place their products.

Yet Memorex suffered financial problems in the early 1970s. In 1970 sales reached $70 million with net income of $3.2 million. The bulk of all sales came from tape media, not from data processing equipment. From 1971 through 1973, Memorex posted net losses on sales and still had substantial debts now due to Bank of America approaching $180 million. Memorex's problems sliced into profits caused by declining demand and increased competition for such items as disk packs and video and computer tapes. Income from media products for the years 1970–1975 dropped by $450 million below expectations. Microfilm products for computers (first shipped in 1970) hardly grew by 1973, achieving less than 10 percent of what Memorex had expected. Industry watchers suspected Memorex had failed to do its homework in conducting even a market survey to gauge demand. Sales of terminals also fell below planned rates by over $40 million in the first three years of the 1970s. Even with its plug-compatible disk drives, Memorex experienced manufacturing problems and declining quality control in 1972 and 1973, concerns that were not resolved until late 1974. In 1972 Memorex brought out two computers, called the Memorex Model 40 and 50; they functioned poorly and had to be withdrawn in July 1973.

The company now found it difficult to raise capital to cover its expenses. Publicity on its manufacturing and quality problems began to take a toll. Criticisms of its accounting practices (such as the way it established subsidiaries) tainted the firm's image. A major problem was the company's long-standing practice of deferring its expenses for the acquisition of leases, research, and development. The issue became so severe that the New York Stock Exchange suspended trading on the company's stock in 1970. The Securities and Exchange Commission (SEC) conducted an investigation. The SEC sued for fraud; the case was settled in 1971. In 1973 the company changed its accounting practices and that year wrote off in excess of $37 million of expenses which it had accumulated on its accounting books.

Such turmoil spelled doom for many executive careers. Between March 1971 and June 1974 nearly all of the company's most senior executives were replaced. The company began to thrive under the new management. In 1979 revenues reached $600 million for data processing products alone, and total worldwide sales climbed to over $700 million. In contrast to the difficult year of 1973, when sales hovered at $173 million, the company was profitable. Sales were based on new disk and tape drives, computer memories, and communications products; almost all of these were aimed at the IBM plug-compatible market. Each time IBM announced a new disk drive, Memorex brought out a ''look-alike.'' Thus, for example, when IBM announced the 3340, Memorex brought out the 3640 in 1976. When IBM introduced the 3350, Memorex followed with the 3650. The company executed the same strategy with tape drives. In 1978 Memorex made public its 3770 disk cache and in the following year came out with the 3652 disk subsystem, one of the first dual-density 3350-type plug-compatible products in the industry.

In the late 1970s Memorex strengthened its hand through acquisitions, or so it seemed at the time. Business Systems Technology, a manufacturer of storage and communications equipment, came into the fold, and in 1978 Memorex and Teijin began a joint venture to produce flexible media in Japan. In 1978 Memorex also acquired Telex Europe. These acquisitions did not help the company, however, and in 1980 it logged its first loss on sales since 1974.

On the legal front Memorex also suffered a reversal. In the late 1970s its suit against IBM went to trial and resulted in a directed verdict in favor of IBM in 1978. Hence, Memorex would have to compete without any help from the courts, and any hope of obtaining badly needed capital through fines that might otherwise have been imposed on IBM in the form of damages disappeared.

In the 1980s the company continued its strategy of the late 1970s but played a far less significant role than it had in the previous decade. In 1981 Burroughs Corporation acquired the company. In the early 1980s Memorex fell behind IBM in the development of new disk products to market against IBM's 3380s. The inability to keep up cost Memorex market share and the opportunity to be more than marginally profitable. Although the firm went into eclipse by 1983, it was nonetheless a $1 billion enterprise, and Burroughs retained hopes that Memorex would expand and at IBM's expense.

For further information, see: Franklin M. Fisher et al., *IBM and the U.S. Data Processing Industry: An Economic History* (New York: Praeger Publishers, 1983); Franklin M. Fisher et al., *Folded, Spindled, and Mutilated: Economic Analysis and U.S. v. IBM* (Cambridge, Mass.: MIT Press, 1983); Stephen T. McClellan, *The Coming Computer Industry Shakeout* (New York: John Wiley & Sons, 1984).

MICROSOFT CORPORATION. This company became the largest supplier of software for microcomputers (also known as personal computers) in the United States. It is one of many examples of data processing companies that grew rapidly from a garage-like existence run by a president that had barely attained adulthood. More closely a success story within the software subindustry, Microsoft represents one of thousands of software houses that emerged in the 1970s and 1980s to supply application programs to the rapidly expanding number of end users within the American economy.

As with many other small, highly successful entrepreneurial operations, the history of Microsoft was closely intertwined with the role of its founder, William (Bill) H. Gates III.** He and his close friend, Paul Allen, had first stumbled across small computers as children, and by the time they graduated from high school, both had considerable experience with small systems. Gates has been called one of the first "hackers" (programmers who break into systems via software). When only fifteen years old he and Allen established a company called Traf-O-Data which offered a service, based on computer chips,* to monitor traffic patterns in Seattle. That year the company's revenues reached $20,000. He next took a year off from high school to write programs for TRW.†

The genesis of Microsoft dates back to February and March 1975 when the two friends worked on a version of BASIC* (the language that would become the most widely used by owners of personal computers). Encouraged by their demonstration of their programming package to MITS, one of the first companies to make micros, Gates left Harvard to set up Microsoft with Allen. At the age of twenty-one, Gates moved the company to Seattle where he spent the next several years writing software for microcomputers. He did so just as this technology began to appear in quantity; thus Gates, in effect, prospered with that segment of the data processing industry as it flourished. The company came into existence in 1975; in 1980 International Business Machines Corporation (IBM)† signed an agreement with the firm to produce the operating system for its planned microproduct. At the time, Microsoft boasted a payroll of thirty-two employees.

In August 1981 IBM announced the Personal Computer (PC) which used Microsoft's operating system. Within three years, the PC dominated the world of microcomputers. Over 2 million devices used Microsoft's operating system, along with other application packages and programming languages, by the mid–1980s. Microsoft's operating system, called MS-DOS, was subsequently licensed to 100 other manufacturers, making it the de facto standard in the industry. This privately held company broke the $100 million barrier in 1984 with profits of approximately $15 million. Gates was now twenty-nine years old and a millionaire. That year the payroll serviced 620 employees. In the mid–1980s Microsoft developed programs to operate on almost every major microcomputer in the industry, including those built by Apple Computer† and Tandy Corporation†/Radio Shack.

The company was successful because it combined vision and good timing with technical expertise and a practical approach. Gates was good at selling his ideas and bringing projects to a successful resolution, and when his company began to grow, he possessed the wisdom to realize that he needed professional managers. Thus, the company moved from a relatively small operation into a fully developed organization. It could now venture into new areas of software. By the mid–1980s Microsoft had established standards which other software houses used to design and produce products.

For further information, see: Paul Freiberger and Michael Swaine, *Fire in the Valley* (Berkeley, Calif.: Osborne/McGraw-Hill, 1984); Robert Levering et al., *The Computer Entrepreneurs* (New York: New American Library, 1984).

MOORE SCHOOL OF ELECTRICAL ENGINEERING. At this school, which is part of the University of Pennsylvania, the first electronic digital computer,* called the ENIAC,* was built in the mid–1940s. During World War II it was one of the world's leading centers for research on computerized equipment. Like other locations armed with engineering expertise, the Moore School turned its attention to war-related projects and initially began construction of a machine to prepare and then print firing tables. The machine was proposed in April 1943, and its construction was authorized by the U.S. Army in June. It was completed by November 1945 and ran its first programs the following month. The machine was dedicated on February 16, 1946, and was moved to the Aberdeen Proving Ground in November of the same year. The school's most important years, insofar as data processing was concerned, were 1943–1946 when it was involved with ENIAC and related work.

Most American radio and electronics firms were located in Philadelphia in the 1920s and 1930s. The Moore School had been heavily involved with electrical engineering and therefore was a natural location for courses taught to participants in war-related projects in 1941 and 1942 for the military. The U.S. Army became impressed with the capabilities of the school in the 1930s, particularly when Herman H. Goldstine,** a scientific adviser to the military, took an interest in the engineering staff there in the early 1940s. John von Neumann,** whom many historians consider the architect of the modern computer, also became interested in the Moore School and frequently advised its staff on the ENIAC and its successor, the EDVAC.* His involvement was important because he influenced many of the government's computer-related decisions during World War II, including endorsement of support for the ENIAC.

The school was founded in 1923 as a result of a bequest made in 1922 by Alfred Fitler Moore at his death. Moore had been a manufacturer of insulated electrical wire and sought an institution for the training of electrical engineers. By the early 1930s it was a fully established institution and interested in differential analyzers. With government support it worked with MIT on the construction of an analog device in the early 1930s. Working with Vannevar

Bush** of MIT, the school built a differential analyzer which it completed in 1935. The school occupied a red-brick building on Walnut and Thirty-third streets. By the time World War II was underway its staff included John W. Mauchly** and J. Presper Eckert,** the chief architects of ENIAC. Research at the school was under the direction of John Grist Brainerd** who negotiated various research agreements with government agencies. In 1940, for example, he obtained a contract with the U.S. Navy to design airborne minesweepers. He played a similar role in negotiating agreements to support the ENIAC and the EDVAC. Yet the Moore School was small in comparison to MIT.

Located in Philadelphia, armed with experience in gaining government contracts, and blessed with talent specifically needed to build computers, the school was well positioned to play an important role. At the start of World War II, it was considered the second best engineering school in the United States after MIT. The dean of the Moore School, Harold Pender, was well connected with the Aberdeen Proving Ground and had negotiated the original contract for the construction of the differential analyzer. His contacts and experiences paved the way for the Aberdeen Proving Ground to work with the Moore School. At the beginning of the war, it took possession of the differential analyzer built in Philadelphia, in accordance with the original agreement made in the early 1930s should war ever break out. The loss of that machine to the Moore School encouraged the staff to want to build another, perhaps better device. The result was ENIAC.

At the end of World War II, while work finished on ENIAC and preliminary efforts had begun on the EDVAC, a problem developed, which eventually eliminated the Moore School as a major research facility for computing. Irven Travis, a Navy veteran, joined the Moore School at the end of the war and soon after became research head. He immediately sought to modify the school's accounting and business practices to bring them in line with those in use at Johns Hopkins University and at MIT. He, together with the dean of the school, believed that no faculty or staff member should benefit financially from work done at the university. That issue became critical in 1946 as various scientists around the country began considering the commercial possibilities of their wartime research. Those who had worked on computers were not immune to such feelings. Travis sought to put all war-time developments under patent in the name of the university. Meanwhile, General Electric (GE)† wanted to negotiate a contract valued at $1 million for research.

In order to win that contract, Travis had to obtain a patent release from all members of the Moore School. When he announced that fact in March 1946, some of the engineers, including Eckert and Mauchly, protested. They felt entitled to patent rights on the ENIAC because they had a signed agreement with the president of the University of Pennsylvania. They were also interested in potential patents that might become viable as a result of work being done on the EDVAC. Travis insisted that they sign a patent release, but they refused. They left the Moore School and established their own firm which, in time, built many computers, including the UNIVAC.* After these two engineers left, other leading researchers on the ENIAC also departed. Within several months, Travis's actions had ended in the breakup of the largest concentration of computer scientists in the world. A few engineers stayed to work on the EDVAC; another group followed Eckert and Mauchly; and a third contingent joined von Neumann at the Institute for Advanced Study at Princeton. The Moore School's computing capability was only a hollow shell by Thanksgiving 1946.

The breakup of the Moore School's computer development projects was a tragic loss, and historians are still in disagreement over why it was allowed to happen. One reason frequently cited is the school's obvious lack of appreciation for the significance of the ENIAC and for computing in general in 1946. Computing was commonly undervalued in many universities, electronics firms, and some government agencies at the time. A second cause cited is, of course, Travis's insistence that engineers relinquish patent rights. Some faculty members thought the entire controversy highlighted the Moore School's insufficient attention to training engineers and excessive attention to research contracts and possible profits. Regardless of the reasons, the net result was that the Moore School never again played an important role in data processing. It attempted to do research and in time installed computers for use by its students, but it never again matched the contributions it made in the 1940s.

For further information, see: Stan Augarten, *Bit by Bit: An Illustrated History of Computers* (New York: Ticknor & Fields, 1984); Herman H. Goldstine, *The Computer from Pascal to von Neumann* (Princeton, N.J.: Princeton University Press, 1972); Margaret Harmon, *Stretching Man's Mind: A History of Data Processing* (New York: Mason/Charter, 1975); Joel Shurkin, *Engines of the Mind: A History of the Computer* (New York: W. W. Norton, 1984); Nancy Stern, *From ENIAC to UNIVAC: An Appraisal of the Eckert-Mauchly Computers* (Bedford, Mass.: Digital Press, 1981).

N

NATIONAL BUREAU OF STANDARDS (NBS). This U.S. government agency encouraged and supported the development of electronic digital computers* in the United States immediately following World War II. As a result, it was closely associated with three of the more important early computers of the era: the UNIVAC,* SEAC,* and SWAC.* NBS was established within the U.S. Department of Commerce in 1901 to provide standards for all kinds of measurements used in the United States. By World War II it was the nation's primary facility for many research projects involving standards in length, mass, time, volume, temperature, light, color, electrical energy, and radioactivity, among the many areas it influenced. Scientists at NBS played important roles in the physical sciences and mathematics in the 1920s and 1930s. It also became interested in computational studies and supported the development of computers, an area of concern which it retains to the present. Its Institute for Computer Science and Technology has responsibility for the bulk of the activities at NBS in the general field of data processing.

The most dramatic example of NBS's role with computers came with the UNIVAC. During World War II, the team of J. Presper Eckert** and John W. Mauchly,** at the Moore School of Electrical Engineering† at the University of Pennsylvania, built the first electronic digital computer, called the ENIAC.* In 1946 they established their own company, originally called the Electronic Control Company and later the Eckert-Mauchly Computer Corporation, for the purpose of building commercially available computers. By this time ENIAC was operational, and its follow-on, the EDVAC,* well underway.

Eckert and Mauchly sought and gained a contract from the U.S. Census Bureau to build that agency a computer. In order to help finance its early stages of development, the NBS (then under the leadership of Edward Uhler Condon, a physicist) participated in supporting feasibility studies while advising the Census Bureau. That contract was signed in September 1946, and, on May 24, 1947, all parties concerned decided to call the new machine the UNIVAC (UNIVersal Automatic Computer). In 1948 NBS also agreed to sponsor the actual construction of the device. Since the financial support offered Eckert and Mauchly did not quite cover all expenses, the two engineers momentarily put aside that project to build the BINAC* for Northrop Aviation.† Having completed that job, the two then refocused attention on the UNIVAC which they completed in March 1951. Although the UNIVAC went on to become one of the best known and successful of the early electronic computers, without the sponsorship of NBS it might not have been built.

Work at NBS on other computer-related projects took place while Eckert and Mauchly were busy with BINAC and UNIVAC. Efforts before World War II set the pace for NBS. Under the directorship of Lyman Briggs, who ran the agency from 1933 to 1945, NBS expanded its activities in the twin areas of mathematics and physics. In 1938, as part of the government's overall program of providing employment during the Depression, NBS hired mathematicians to develop new mathematical tables. By the time Briggs retired, this work had grown so extensively that NBS had to establish the National Applied Mathematics Laboratories as a division. The U.S. Navy strongly encouraged the creation of this new division in an attempt to pool the various agencies' expertise in computational devices and related technologies which it wanted to use in the postwar period. Condon, Briggs's successor, was committed to the use of his agency's resources and power to foster peaceful uses of science and, specifically, computers. Ironically, as during World War II, it was the military community's enthusiastic support of computers that facilitated Condon's task afterwards. Nonetheless, his creation of various laboratories in the late 1940s within NBS permitted numerous peacetime projects involving computers, which also gained the military's approval and support.

The new mathematics laboratory was a case in point. It was divided into four groups: Computation Laboratory, Machine Development Laboratory, Statistical Engineering Laboratory, and Institute for Numerical Analysis. The last was established in Los Angeles near the University of California, and the others were headquartered in the Washington, D.C., area. The Applied Mathematics Laboratories were very aggressive in obtaining UNIVAC because the U.S. Air Force, which was then eager to use computers in operations research, could not wait until others (such as the Institute for Advanced Study) completed their machines. Thus, it came to NBS for help in 1948, hoping to get a computer quickly. Condon was then trying to make his bureau the central repository of information and leadership for all of the government's computer projects and thus welcomed the opportunity to help the Air Force. In these years, NBS discovered that interest in computers was so widely dispersed throughout the government that it could never become the central focus of all computer-related activities. Nonetheless, NBS constructed or contracted for a number of important machines.

In the case of the U.S. Air Force, NBS built what became known as the SEAC

(Standards Eastern Automatic Computer) but was initially called the National Bureau of Standards Interim Computer. It was interim until UNIVAC could become available. SEAC became operational in May 1950. NBS also constructed the SWAC, more formally known as the Standards Western Automatic Computer, which was used at the Institute for Numerical Analysis in Los Angeles. The DYSEAC, also built by NBS, was the Second Standards Electronic Automatic Computer. DYSEAC was built for the U.S. Army Signal Corps and became operational in April 1954. During the course of building these machines in the late 1940s, scientists and engineers at NBS advised various agencies on the use and construction of computers, including the Census Bureau. It worked closely with the Office of Naval Research (ONR),† one of the most important government supporters of university research in the physical sciences and in mathematics. Between the ONR and the NBS, the government gave crucial support to research and development of electronic computers in the years immediately following World War II. During this period electronics firms were either just being formed to build such machines and were thus still starved for capital (such as Eckert's and Mauchly's firm), or had not yet made the necessary commitment to this new field of electronics (General Electric,† for instance).

For further information, see: Herman H. Goldstine, *The Computer from Pascal to von Neumann* (Princeton, N.J.: Princeton University Press, 1972); Nancy Stern, *From ENIAC to UNIVAC: An Appraisal of the Eckert-Mauchly Computers* (Bedford, Mass.: Digital Press, 1981); Michael R. Williams, *A History of Computing Technology* (Englewood Cliffs, N.J.: Prentice-Hall, 1985).

NATIONAL CASH REGISTER COMPANY (NCR). Although not a producer of data processing equipment until after World War I, NCR employed Thomas J. Watson,** later founder of International Business Machine Corporation (IBM),† and through him influenced the management style of Watson's firm. During the interwar years, NCR flirted with data processing, selling accounting and tabulating equipment. Yet during its first forty years its major product was the cash register. By the early 1950s it was moving into the area of electronics and had developed computers, and by 1979 it had become a major vendor.

Many of NCR's business practices became IBM's and later those of the entire industry. NCR, or as it was known in the nineteenth century, "the Cash," was the product of a highly creative and tough-minded businessman, John Henry Patterson (1844–1922). In addition to developing NCR into a major organization, Patterson deeply influenced other important companies, many of which were high tech for their day (automobiles, printing equipment, etc.), because he trained and fired executives who went on to run other companies as chairmen, presidents, and senior vice-presidents.

While running a dry goods operation in Ohio, Patterson discovered that the business lost money because clerks kept receipts. Hearing of a device (later called a cash register), which could prevent such thievery, he ordered two at $50 each from a James and John Ritty. The Rittys had a company originally called the National Cash Register and in 1882, after selling off part interest in it, renamed it the National Manufacturing Company. Meanwhile, Patterson had such great success with his registers that he bought stock in the company, joined the board of directors, and proposed that these devices be sold nationwide with creative new marketing programs. When his ideas encountered opposition, he bought additional stock and took control of the company in 1884. Once its president, he renamed it the National Cash Register, thoroughly reorganizing its sales and manufacturing operations.

During the first year under his control (1885), it has been estimated that the company sold several dozen cash registers generating a gross income of about $1,000. By 1913 Patterson had sold more than 1.2 million NCR cash registers, and some 13,000 were being manufactured each month. By the end of 1913 sales had grown to $21 million yearly. Although his products were no better than others that were available, his sales organization was far more efficient than that of any of his competitors.

Patterson created a corporate culture that idolized salesmen. As IBM was to do years later, he paid his salesmen well, giving them more attention than manufacturing. In effect, he reversed the emphasis of many American companies in the late nineteenth century. His salesmen were expected to dress well, yet conservatively, and to work hard, embrace a business ethic, and be motivated to be efficient and productive for the company. Patterson was one of the first American executives to establish the practice of assigning specific territories to salesmen which other representatives of the company were not to breach. (This was yet another reversal of standard policy of the day.) Patterson issued quotas and rewarded performance with bonuses and good salaries. Those who made their sales quotas became members of the Hundred Point Club. (Watson did the same thing at IBM and called it the Hundred Percent Club—an institution that is still preserved today at IBM.) Patterson paid salesmen better than most of his competitors, thus motivating them to do well. In addition, NCR's advertising campaigns were more expensive and original than those of his competitors. He instituted a series of measures which today are commonly called account control. For example, he insisted that salesmen maintain contact with their customers even after products had been sold so that additional opportunities could be quickly identified and seized on. In 1894 Patterson established a training school for salesmen to learn about the company, the product line, and methods of selling. Many of the principles established in NCR's training program today characterize IBM's principles for newly hired sales personnel.

Patterson was the author of the phrase "Think" which for many years was the single word most widely associated with IBM. Patterson's legacy lives on in many other ways as well. The use of five cards on which all salesmen had the phrases necessary to pitch the case for a register, in modified form, resulted in the teaching of various types of scripted planned sales calls in IBM. Patterson also initiated the use of large sheets of paper for presentations. Today all companies that either sell or use data processing are probably familiar with the "flip charts" which Watson and IBM made as common as the use of white shirts and dark suits—also part of NCR's nineteenth-century management practices.

Thomas J. Watson joined NCR in 1895 at the age of twenty-one. He was quickly molded into the image of an "NCR man," and his sales career took off. He rose through the ranks to become a branch manager in Rochester, New York. After successfully making that territory lucrative, Patterson brought him into his inner circle of executives in Ohio where he fought competition in the rough and tumble fashion of the early 1900s on a national scale. Watson's actions finally led to a conviction for antitrust activities in 1913, although his sentence was commuted. Meanwhile in April 1914, following a fallout with Patterson (which was a common occurrence with many of the NCR executive managers), Watson left the firm. He swore to run a business that would be bigger than Patterson's—and he would run that business the only way he knew how, in the manner common at NCR.

In 1922 Patterson turned over the reins of power to his son Frederick. The organization was strong and healthy, characterized by an efficient sales force and a clear opportunity to dominate the cash register business. Its great liability was its lack of executive talent, for turnover at that level had been great for many years. The younger Patterson was unable to decide as quickly and as well as his father or to provide the elder's creative spark. Throughout the 1920s the younger Patterson resisted his executives' suggestions that the company move quickly into the rapidly growing market for accounting and tabulating equipment. Already Herman Hollerith** had paved the way, and other firms such as the Powers Accounting Machine Corporation† were growing as a consequence. NCR had an established customer base to which it could have sold accounting equipment. At the end of the decade the decision was made to get into these markets.

Regardless of the internal struggle to define what business NCR should be in, revenues increased. In 1921 sales totaled $29 million, and in 1925 they had grown to $45 million. Profits in the same period increased from $2.8 million to $7.8 million—a healthy growth by any business standard. Yet during the second half of the decade, sales flattened as competitors made inroads into cash register sales and new products were not forthcoming from the company. If we use 1928 as a benchmark, out of the top five business machine firms in the United States, NCR had dipped to second behind Remington Rand† in revenues and second to Burroughs† in earnings.

Recognizing the need to broaden its product offerings, NCR decided to offer adding and accounting machines to its customers. Nonetheless, it entered the era of the Great Depression weakened. Its Class 2000 accounting machines could not halt the decline in sales. In 1931 the company was on the brink of bankruptcy, and so the younger Patterson turned over control of the company to Edward Deeds who set out to rebuild the corporation. In 1943 NCR purchased the Allen-Wales Adding Machine Company. Earlier, under Deeds' direction, NCR had begun the process of developing some in-house expertise in electronics and initiated a modest program of research and development. During World War II it built its first computer-like device for the government to calculate bombing navigational data. Following the war, NCR continued to experiment in the development of electronically based products.

Meanwhile, business boomed after the war. By 1953 annual sales approached $260 million. Interest in computing had also grown. That year NCR bought Computer Research Corporation (CRC)† which enabled the company to acquire an even broader base of technical knowledge in electronics. CRC had built a small general-purpose computer for the military and had worked out some processes for manufacturing multiple copies of the device. In addition to the purchase price of $1 million, NCR invested an additional $4 to $5 million in CRC within the next three years in order to allow the company to enter the computer marketplace. Then in 1953 NCR announced the CRC 102D computer which it intended to sell to commercial customers primarily for scientific work. Development work on a subsequent machine (called the 303) was abandoned in 1956 because of the 303's reliance on older technology, which was now regarded as not competitive in a rapidly changing computer marketplace.

Perhaps because of the experience with the 303, NCR's role in the computer industry remained conservative during the mid–1950s. Data processing revenues from U.S. sales declined from $3.1 million in 1954 to $211,000 the next year and went back up but to the still low figure of $308,000 in 1958. Despite the sharp decline in sales, NCR did announce additional computer-related products. For example, in 1957 it announced the 304 which was a computer based on solid-state technology, and eventually the company sold thirty-three of these general-purpose computers. In 1960 came the 310 computer, which was actually manufactured by Control Data Corporation [CDC]† for NCR. That same year NCR opened up data centers to sell computing services. Commonly used business applications run on its systems included sales analysis, inventory reporting, and payroll. The service bureau segment of the business grew rapidly and within two years was handling several million sets of transactions monthly. More products appeared simultaneously: the 390 computer, also in 1960, the 315 computer in 1962 to replace the 304, and a variety of peripheral devices. The 304 sold nearly 700 copies.

NCR also announced Card Random Access Memory (CRAM) for use with the 315 to provide magnetic storage, but the product failed to catch on. By the early 1960s, NCR had a group of commercial computers with peripherals and software ranging in cost from $5,000 a month leased to $12,000. The larger machine (315) could use COBOL* and FORTRAN.* Sales of the 315 helped the company dramatically. NCR's revenues for EDP products went from its low of $308,000 in 1958 to nearly $31 million by the end of 1963. Many sales of computer-related products were to long-established customers of either cash registers or accounting equipment who now needed additional computing capability. Because of its large customer base, many industry analysts felt that NCR had a real chance to capture a significant portion of the data processing marketplace.

Already in 1964, data processing revenues were contributing 13 percent of all sales to the company, or $46.3 million. Yet it continued to focus considerable attention on cash registers and accounting machines. NCR decided to sell products both in the data processing market and in its old traditional customer set, merging technologies where it made sense. As a consequence, the company split its resources between the two markets: the traditional market for cash registers and accounting devices, and the computer market. This decision was made at a time when some business machine companies, such as IBM, were concluding that in order to be competitive in data processing, an organization would have to commit its full resources to it. Already by the mid–1950s, for example, IBM had recognized that the traditional accounting machines were rapidly becoming obsolete and that all of its resources had to be dedicated to data processing and computers. The impact of NCR's split strategy became evident by the mid–1960s. Using 1965's statistics, in order of gross revenues earned for computer sales, NCR trailed a distant fifth behind General Electric (GE),† IBM, Radio Corporation of America (RCA),† and Sperry Rand†. Yet it was considered one of the "Seven Dwarfs" of the industry when compared to IBM with its rapidly growing influence and strength. In terms of market share, NCR was eighth, with 2.9 percent of the market. IBM had 65.3 percent, followed by Sperry Rand at 12.1 percent and Control Data Corporation at 5.4 percent.

NCR's computer marketplace was solidly in banking and retailing, which represented many of its customers from the precomputer era. Contacts in this industry virtually assured NCR that it would sell some computer products. Throughout the 1960s NCR continued to introduce retail terminals and cash registers, some of which could be linked to computers.

Its technological innovations during the 1960s were conservative. Thus, while its giant competitor IBM took the bold step of announcing the System 360* family of computers, thereby revolutionizing computer technology and causing the industry to explode with growth, NCR only enhanced existing products. Between 1964 and 1968, NCR's second-generation products acquired Rod Memory Computer (RMC) features for the 315. In 1966 multiprogramming for this computer became available. Finally, in 1968 NCR announced a third-generation computer, called the Century Series, with various model sizes to compete with the System/360 which had been announced nearly four years earlier. NCR's third-generation product line allowed the company to market devices to a broader customer set and for a variety of business applications. Orders for the new series pushed sales forward, up 98 percent in 1968 over 1967's order rate.

The data centers were also doing well, expanding throughout the 1960s and totaling sixty-nine by 1968. These centers not only generated revenue for NCR, but also allowed the company's sales force to sell computers when customers sought to bring their computing back in-house into their own data centers.

By 1970 the leading manufacturing companies in the data processing industry, as well as many of its most sophisticated users, were moving in the direction of online computing. Products to support such data processing, as opposed to the more widely used techniques of batch processing, became the norm during the 1970s. NCR, however, had difficulty taking advantage of new conditions calling for additional online products (terminals and software, for example). Consequently, it lost market share. By 1969 the problem had become evident to NCR's top management, which made a significant commitment of resources to redress the problem. During the 1970s new products, particularly terminals, insured NCR's firm commitment to the data processing industry and a continuous revenue stream.

One of the first signs of revival came in 1970 when NCR introduced an industry terminal for use in retail operations. The 280 reflected the kinds of technology many vendors would offer in the 1970s. It was an intelligent terminal, had input/output attachments, could store data, and had the capability of communicating to a mainframe computer. Its software also allowed it to work with other NCR products. In 1971 NCR introduced the 270 Financial System, providing banks with the kinds of features available with the 280 to conduct normal banking transactions. Such products reversed NCR's trend of declining profits which it had experienced between 1969 and 1972. In 1969 profits had been $30 million, but by the end of 1972 management reported a net loss of nearly $60 million. In 1973 revenues grew 17 percent to $1.8 billion with profits of $74 million. Sales of all products (data processing and otherwise) were almost entirely in a modern product line and almost exclusively electronic. Mechanical devices now made only a minor contribution to profits.

Once again NCR was dominated by a creative executive. William S. Anderson became president in May 1972. He had held various executive positions within the company and understood its situation well. He became chief executive officer in July 1973. During his early days as head of the company, he initiated a major reorganization of NCR's manufacturing, development, marketing, and administrative functions. The company elected not to introduce radically new products and technology during this period, building instead on its announced third-generation family of computers, adding new models to the Century series. The data centers grew to over eighty by the early 1970s and expanded their services. The company did introduce additional terminals to meet the requirements of additional industries, covering a large number of major business applications.

In 1976 NCR finally introduced a whole new series of computers and related devices. Additional product announcements and enhancements were made in

1977, taking advantage of newer technology and manufacturing methods. The company was now using its considerable experience with online systems gained from running data centers for a variety of products. Thus, by 1979 its broad experience with distributed data processing insured that the company would enjoy a current product line to generate revenue.

From 1974, when NCR's revenues from computers were $385 million, sales grew to $838 million in 1979. Terminal sales during the same period rose from $295 million to $959 million. NCR's new computer system, the 8000 series, announced in 1976, was well received, and for the rest of the decade, enhancements were announced. NCR entered the 1980s as a full-service, broad-based data processing vendor capable of shipping computers, their peripherals, terminals of a large variety, software, and related services. It was now the third largest supplier of sales and rentals within the industry as measured by revenues across the entire product line. It had become a fully integrated vendor within the data processing industry.

For further information, see: S. Crowther, *John H. Patterson, Pioneer in Industrial Welfare* (Garden City, N.Y.: Doubleday, Page, 1923); F. M. Fisher et al., *IBM and the U.S. Data Processing Industry: An Economic History* (New York: Praeger Publishers, 1983); R. W. Johnson and R. W. Lynch, *The Sales Strategy of John H. Patterson, Founder of the National Cash Register Company* (Chicago: Dartneli Corporation, 1932); I. F. Marcosson, *Colonel Deeds: Industrial Builder* (New York: Dodd, Mead, 1947) and his *Wherever Men Trade* (New York: Dodd, Mead, 1948); Robert Sobel, *IBM: Colossus in Transition* (New York: Times Books, 1981); S. N. Whitney, *Antitrust Policies* (New York: Twentieth Century Fund, 1958).

NATIONAL COMPUTER CONFERENCE (NCC). The NCC is the largest annual computer show within the data processing industry and has been so since its inception in 1973. This event always consists of exhibits by major vendors within the industry and, in recent times, many smaller organizations. It features lectures and seminars on all aspects of data processing from technical issues to reviews of products, from concerns of management to controversies affecting the entire industry. Attendance has always been very high, and in 1985 it approached 80,000.

The first national show was held in New York in June 1973. That year International Business Machines Corporation (IBM)† exhibited its small System/3 Model 10 and its intermediate-sized computer, the 370/145. The keynote address, delivered by U.S. Senator Philip Hart (D-Michigan), concerned the danger of government regulations of the industry. The following year in Chicago, Vice-President Gerald Ford talked about the issue of privacy of information and its relationship to the data processing industry. The 1975 session at Anaheim, California, was the most widely attended thus far, with nearly 32,000 people present as many vendors introduced new products. Some 278 companies had exhibits at this fair. During the week of this particular conference, the U.S.

government filed its antitrust suit against the IBM Corporation, accusing it of monopolizing the industry. (Over a decade later, the government would drop its suit for lack of a sufficient case.) In 1976 in New York, Telenet Communications Corporation demonstrated a packet-switching network supporting sixteen different companies. This conference had the largest number of exhibits ever of terminals. In 1977 in Dallas, although not allowed to formally demonstrate microcomputers at the show, vendors first exposed the industry to such products in large quantities in nearby hotel rooms. By the time NCC programs were being put together for 1983, microcomputer manufacturers had become an important group of exhibitors. Back at Anaheim in 1978, 51,000 members of the data processing industry saw a raft of new products exhibited, particularly memories for IBM's 3030 computers and for teleprocessing.

At the conference in 1979 (in New York) Univac unveiled its 1100/60 computer with which it hoped to challenge IBM's 4300 series. By the end of that year, nearly a dozen companies announced products to compete with one of IBM's most successful computers (4300s). In 1980 for a third time Anaheim became the locale for an NCC and yet again in 1983. The shows became larger during the 1980s as the number of vendors and attendees increased. The number of presentations also rose. In 1985, for example, eighty-four technical sessions were scheduled along with twenty-seven seminars on professional development. More than 600 vendors registered to exhibit their wares. The keynote speaker that year, retired U.S. Navy Admiral Bobby R. Inman, then chairman, president, and chief executive officer of Microelectronics and Computer Technology Corporation (MCC), focused his attention on how to encourage research on computers that could then be converted into marketable products.

For further information, see: Paul Korzeniowski, "NCC Past to Present: A Barometer of Industry Progress," *Computerworld*, July 8, 1985, Preview/8.

NATIONAL RESEARCH DEVELOPMENT CORPORATION (NRDC). NRDC was established in Great Britain to promote the development of British computers during the 1950s. It was founded in July 1949 in order to sponsor the construction of computers based on patents issued on devices built at the University of Manchester. At the time, the British Ministry of Supply held title to them and turned them over to the NRDC. The NRDC also sought to foster the development of a competitive British computer industry that could rival the one just emerging in the United States.

To help meet that objective, the NRDC established the Advisory Panel on Electronic Computers. Its first and only meeting was heavily attended by members of all the important electronics and punched card equipment firms in Great Britain on December 14, 1949. The committee failed to thrive largely because of the hostility of companies unwilling to take their direction from the NRDC. Yet the NRDC proved an effective influence on companies in their early efforts with computers.

The agency accomplished this purpose by subsidizing research which it believed to be beneficial to the British nation. For example, in 1950 it supported the development of the 401 by Elliott Brothers. The following year the NRDC underwrote some of the expenses for the Mark I Star made by Ferranti Ltd.†, and in 1954, for that company's PEGASUS.* EMI Ltd. found the NRDC willing to fund development of office-based computers in 1955. In 1958 the NRDC sponsored the installation of a transistorized computer at Austin Longbridge works. This last computer was interesting in that the device served as the prototype of the EMIDEC 1100 computer (first shipped in 1959) and the subsequent EMI 2400. The latter was built to the specifications established within the NRDC. To insure competitive tape drive subsystems in Great Britain, the NRDC began supporting this kind of development beginning in June 1953.

The NRDC also took measures to patent the results of research it sponsored, particularly in the early 1950s. By 1956 it was managing over 730 patents generated by 201 inventions. Manchester University invented eighty-one items, the National Physical Laboratory (NPL) forty-three, the Ministry of Supply thirty-six, and Ferranti sixteen. The NRDC wisely licensed these patents, using the resulting revenues to sponsor additional research.

During the 1950s the NRDC developed some in-house expertise in the field of computing. This is no better illustrated than in the person of Christopher Stratchey** who programmed the St. Lawrence Seaway Project (1952–1953) and advised on the construction of the Elliott 401 (1953–1954), among other machines. This organization also sponsored the establishment of an institution later known as the British Computer Society during the mid–1950s as yet another means of proliferating experience with data processing.

The early 1950s witnessed basic technological innovations, leading to the widespread use of computers by 1960 in both Europe and the United States. In the case of the British, the support provided for the development of computers by the NRDC was crucial to the survival of any chance that Great Britain would play in the field of computers. The academic community, home of most research on data processing, saw little commercial value to this technology in the late 1940s whereas in the United States many companies were founded for the purpose of selling processors. The NRDC's efforts in Britain to establish a computer industry in the British isles was a solid and even visionary step for the early 1950s. Simply calculating the number of machines installed in Great Britain suggests the success of the NRDC. In the late 1940s, there were fewer than ten digital electronic computers* in Great Britain; by 1960 that number had grown to nearly 220, over 185 of which had been manufactured commercially by nine British firms. By 1970 the 220 had grown to 6,500, of which nearly half were constructed in Great Britain.

For further information, see: Simon Lavington, *Early British Computers* (Bedford, Mass.: Digital Press, 1980).

NBS. See NATIONAL BUREAU OF STANDARDS

NCC. See NATIONAL COMPUTER CONFERENCE

NCR. See NATIONAL CASH REGISTER COMPANY

NORTHERN TELECOM, LTD. (NTL). By the early 1980s this Canadian-based company became the second largest manufacturer of telephonic switching equipment in the world and an important vendor of other telecommunications products. NTL's products are generally more advanced technically than those manufactured by telephone and telecommunications companies in the United States (where it is called NTI). This was especially so between 1976 and 1985.

The company originated in the 1880s when the Bell Telephone Company of Canada established a manufacturing facility in Montreal. During the next fifty years, this facility produced equipment solely for the use of Bell Canada, relying on procedures and technology exported by Western Electric. In 1914 the Imperial Wire and Cable Company and the Northern Electric Manufacturing Company merged to create the Northern Electric Company. This new firm was owned 50 percent by Bell Canada and 44 percent by the Western Electric Company in the United States.

In 1956 the U.S. Department of Justice and both Western Electric and American Telephone and Telegraph (AT&T)† signed a consent decree closing out a complaint made by the U.S. government in 1949 over perceived monopolistic practices on the part of the giant telephone empire. This decree required that Western Electric sever its ties to Northern Electric. Although not well prepared to backfill the services provided by Western Electric, the Canadian manufacturer ramped up its skills. In 1976 it changed its name to Northern Telecom, still a subsidiary of Bell Canada, but launched into the future with its own identity. In the twenty years following the Consent Decree, NTL acquired technologies that were essentially developed outside the firm. Sales remained overwhelmingly in Canada, with some exports to Turkey and the United States. By the end of the decade, Bell Canada's share of ownership dropped to 53 percent while common shares were sold largely to buyers in the United States. By the mid–1980s, some 80 percent of all NTL's products were developed in labs which it ran or were owned by Bell Canada (called Bell Northern Research or simply BNR). The company also marketed products in ninety countries.

By 1979 Northern Telecom had become the second largest producer of telecommunications equipment in North America after AT&T/Western Electric. That year about 40 percent of its revenues came from sales in the United States; this total grew to 60 percent by 1985. The growth of this company could be explained primarily by its use of more modern technologies than AT&T used in the United States. Beginning in 1976, it introduced switching equipment (used to transfer hundreds or hundreds of thousands of telephone calls) that were based on contemporary digital technology, while AT&T still relied on less sophisticated equipment which it was amortizing over long periods of time (in some cases forty years). By acknowledging that the technologies of telecommunications and data processing were merging, NTL was able to make products that were more price competitive than AT&T's. After the breakup of the U.S. telephone empire in the early 1980s, this circumstance allowed NTL to sell better quality products less expensively to area telephone companies and for use by companies in private networks. The key was using digital switching systems. The first year such technology appeared (1976), called Digital World Systems, NTL introduced computers as part of that product set. Its switches, called the Digital Multiplex System (DMS), constituted a family of products ranging in size which NTL continuously enhanced with innovations throughout the 1980s.

Meanwhile, NTL expanded heavily and successfully into the United States. In 1978 it acquired three firms as part of this effort. The first, Danray, built auxiliary data exchange equipment that made it possible to connect computers, terminals, word processing equipment, and other data processing hardware into an internal telephone system within a customer's company. The second, called Data 100, founded in 1968, built terminals and small computers used for remote job entry (RJE), local and remote management of files, and other minicomputer functions. The third, known as Sycor and established in 1967, sold terminals and distributed processing products with initial shipments in 1969. By 1977 its products were strategically important to NTL which viewed this acquisition as a means of doing remote processing (using the Sycor 445 system) and SYCORLINK which also made it possible for multiple computers to communicate with each other. The following year, NTL formed Northern Telecom Systems Corporation (NTSC) to handle electronic office equipment sales. Both Data 100 and Sycor were merged into this organization, and in 1979 NTSC reported revenues of $349.8 million (18.4 percent of all corporate sales). By 1979 NTL also owned 21.9 percent of Intersil, which built integrated circuits and memory compatible with International Business Machines Corporation's (IBM's) computers. Thus, NTL was in a good position within the U.S. data processing and telecommunications industries.

Facilities also grew from twelve plants in 1970 to forty-six by 1985. Revenues in U.S. dollar terms went from $436 million in 1970 to $2.6 billion in 1983, and for the entire company in 1983 to $3.3 billion, some $5 billion in 1985. During the 1970s through 1982, growth in the United States proved extensive. In 1971 sales had been only $40 million but by 1985 exceeded $2 billion. By 1983 there were fourteen manufacturing facilities, fourteen research and development locations, and more than 100 sales and service offices just in the United States. Its Data Management Switches (DMS) were being built in Raleigh, North Carolina, there were products in other markets all over the United States, and sales were recorded in all fifty states. Small PBX switches for buildings made inroads, along with products that combined personal computers and telephones into single products, and office automation for word processing and departmental data processing. By the start of 1986, the number of employees in the United States reached 17,000 and nearly 40,000 worldwide. The U.S. subsidiary, known as Northern Telecom Incorporated (NTI), was headquartered in Nashville, Tennessee, under the presidency of Desmond F. Hudson. As of late 1985, the balance of importance had shifted from NTL to NTI in terms of sales generated, although NTL still dominated NTI and Canadian managers retained the key jobs. In 1985, as measured in volume of sales, NTI was the sixth largest manufacturer of telecommunications equipment in the world and the second largest in the United States.

For further information, see: Pamela Archbold and John Verity, "The DATAMATION 100," *Datamation*, June 1, 1985, passim, p. 84; Franklin M. Fisher et al., *IBM and the U.S. Data Processing Industry: An Economic History* (New York: Praeger Publishers, 1983).

NORTHROP AVIATION. Northrop was one of the first private companies to recognize the potential value of computers following World War II. It supported the development of the BINAC,* a computer constructed by J. Presper Eckert** and John W. Mauchly,** the builders of the ENIAC,* EDVAC,* and, after BINAC, the UNIVAC.* Although it thus established an early interest in computing which it retains to this day, its involvement with the BINAC proved the most interesting aspect of its role in the data processing industry.

Eckert and Mauchly, searching for funding with which to continue their work on what ultimately became UNIVAC, contracted with Northrop (then known as the Northrop Aircraft Company) in October 1947 to build a small computer, the BINAC. Northrop wanted the machine to help manage a long-range guided missile it was building for the U.S. Air Force called the "Snark." A computer offered the possible solution to its need for in-flight navigational control. It wanted a computer small enough that it could be taken aboard an aircraft. Northrop took possession of the machine in August 1949 and at a cost of $100,000; it had cost the manufacturer, Eckert-Mauchly Computer Corporation, $278,000.

Northrop's engineers were disappointed with the computer, particularly with the way it operated and its reliability. But the machine had not been designed for transport from the East Coast to California, which accounted for many technical difficulties. Some engineers at the company had hoped to build their own computer, but it became impossible to do once the firm contracted for the BINAC. Although they made the BINAC operational, this, the first stored-program electronic digital computer* built in the United States, was not used for the missile project; instead, gyroscopes were employed. The project taught Northrop's engineers much about computing, giving them skills they would use in other computer-based development of military hardware during the 1950s.

For further information, see: Nancy Stern, *From ENIAC to UNIVAC: An Appraisal of the Eckert-Mauchly Computers* (Bedford, Mass.: Digital Press, 1981).

NRDC. See NATIONAL RESEARCH DEVELOPMENT CORPORATION

NTL. See NORTHERN TELECOM, LTD.

O

OFFICE OF NAVAL RESEARCH (ONR). ONR, an agency of the U.S. Department of the Navy, was a major source of funding for important projects involving the initial development of computers between the end of World War II and the early 1950s. It supported the study of numerical analysis in the new field of operations research at American universities, and it also advised government agencies about their own scientific programs. The ONR was the primary funding agency for the WHIRLWIND* computer developed at MIT in the late 1940s and early 1950s. Mina Rees,** a mathematician who encouraged many government departments to support basic and applied scientific research, worked for ONR. Until the establishment of the National Science Foundation in 1950, ONR was the largest supporter of scientific research in American universities.

Although the full history of the ONR has yet to be written, its role in data processing has been well defined. The agency was established in 1946 in order to maintain the momentum of scientific research which had begun during World War II. It was chartered to support research of benefit to the U.S. Navy and to national defense. Almost from its inception, it took an interest in computers. Mina Rees, within the Mathematics Branch, was a particular supporter of this field of research. Thus, for example, in 1947 ONR participated in the establishment of the Association for Computing Machinery (ACM)† and advised the National Bureau of Standards (NBS)† on its forays into computing. ONR's primary computational interest centered on numerical analysis "that would affect the design of computers and be responsive to the mathematical difficulties that could develop with the use . . . of the powerful computers." This quote from Mina Rees, made in 1982 to describe ONR's focus, captures the emphasis she placed on mathematical issues rather than engineering. That perspective later created friction when the ONR questioned the value of what Jay W. Forrester** and his staff were pursuing as they constructed WHIRLWIND, nearly causing her agency to cancel the project.

Early programs included a contract signed with Engineering Research Associates (ERA)† which led to the publication of an important study on the state-of-the-art of computing as of 1950. ONR also financed various symposia to help disseminate information. It supported the establishment of several laboratories; the most important for computing was the Institute for Numerical Analysis (INA) at the University of California at Los Angeles (UCLA). The ONR also helped NBS to advise the Census Bureau when Census sought to acquire a computer in the late 1940s. Many government agencies were acquiring such machines and turned to the ONR for consultation.

In the form of consultation and funding, the ONR participated in the construction of the SEAC* (1948–1950) and in the development of the Institute for Advanced Study (IAS) Computer* during the late 1940s. Its most famous project was WHIRLWIND which it supported from its inception in 1947 until its completion in the early 1950s. WHIRLWIND became the largest, most expensive digital computer* of the early years of computing. ONR was a reluctant supporter during those years when progress on the machine seemed too slow.

In addition to work on WHIRLWIND and the IAS Computer—two very important machines of the period—ONR also supported work done on the X-RAC, CALDIC, and Logistics Computer. The X-RAC was the only analog computer* ONR supported. It was constructed at Alabama Polytechnic Institute and was finished at Pennsylvania University. Raymond Pepinsky managed the project between May 1947 and early 1950. The CALDIC was constructed by Paul Morton at the University of California at Berkeley in the late 1940s and went "live" in 1951. The Logistics Computer was put together for use by the U.S. Navy to be employed in logistics research using linear programming. Primary work on the machine was done under contract with engineers at George Washington University.

Although ONR continued to support data processing research projects in the 1950s, its great era had ended by 1954. First, Rees went back to Hunter College in New York City in 1953. Second, the previous computer projects ended as their machines went into production. A third factor was the introduction of commercially built machines by such companies as Univac (later Sperry Rand)† and International Business Machines Corporation (IBM).† However, the ONR had served a critical need in its time as the single most important source of funding for basic research (particularly in mathematics) in the United States. Without ONR it would have been difficult for the IAS Computer to have been built and impossible for WHIRLWIND. Its advice to government agencies, including that which it gave to NBS and the Census Bureau, not to mention various military agencies, encouraged the successful adoption of computers and the advancement of the general field of mathematics.

For further information, see: J. H. Curtiss, "The Institute for Numerical Analysis of the National Bureau of Standards," *Monthly Research Report of the Office of Naval Research, May 1951* (Washington, D.C.: U.S. Department of the Navy, 1951): 8–17; Engineering Research Associates, Inc. (W. W. Stifler, Jr., ed.), *High Speed Computing Devices* (New York: McGraw-Hill, 1950, reprinted, Cambridge, Mass.: MIT Press, 1984); Office of Naval Research, *A Survey of Large Scale Digital Computers and Computer Projects* (Washington, D.C.: U.S. Department of the Navy, 1950); Kent C. Redmond and Thomas M. Smith, *Project WHIRLWIND: The History of a Pioneer Computer* (Bedford, Mass.: Digital Press, 1980); Mina Rees, "The Computing Program of the Office of Naval Research, 1946–1953," *Annals of the History of Computing* 4, no. 2 (April 1982): 102–120; J. J. Wolf, "The Office of Naval Research Relay Computer," *Mathematical Tables and Other Aids to Computation* 6, no. 40 (1952): 207–212.

ONR. See OFFICE OF NAVAL RESEARCH

OSBORNE COMPUTER COMPANY. Osborne was one of the earliest companies to make microcomputers and the very first to manufacture and sell a portable version. Along with the Apple Computer Company,† it helped launch major growth in the use of computers in the United States in the early 1980s. Its founder, Adam Osborne,** was part of a new generation of entrepreneurs from California's Silicon Valley, breaking new ground within the industry and bringing small high-technology-based companies to an end.

Osborne, an engineer, published a book, *An Introduction to Microcomputers* in 1975, which was an extension of a consulting job he had done earlier. Through *An Introduction* Osborne foresaw the need for a company to produce a low-cost, portable microcomputer. As a member of the Homebrew Computer Club†—then a hotbed of micro enthusiasts at Stanford University—he had met its acknowledged leader Lee Felsenstein. Osborne commissioned Felsenstein to design a micro for him, and in March 1981 this machine was introduced as the Osborne 1 at the West Coast Computer Fair. The product and his new company (officially established that year) were instant successes.

Osborne produced the only portable, commercially available microcomputer at that time. It weighed 28 pounds and could fit on a seat and thus was marketed as a device that could be taken from work to home and back. It had a five-inch screen which, although a disadvantage when one considered that most cathode ray terminals (CRTs) then and later were at least twice the size of his, nonetheless was new and different. He also offered software for the machine. These included the CP/M operating system which was an industry standard at the time, WordStar which made the machine a word processor, SuperCalc to allow for spreadsheet applications, and C-BASIC and M-BASIC for programming. He priced the entire system of hardware and software at $1,795—nearly a third less than the Apple II. Because it was the most attractive microcomputer then available, the Osborne Computer Company received considerable attention within the industry and from the American business community generally. In 1981 sales totaled $6 million, and the next year, $70 million. Employment grew to 1,000 people at its peak, and 10,000 units were sold each month. During its heyday in 1982 the company raised between $34 and $40 million in capital. Adam Osborne hired Robert Jaunich, an experienced manager, to run this rapidly growing organization.

But growth had its problems. In addition to the problem of how to manage such expansion, there were others involving profitability, management of people, manufacturing, product development, and, ultimately, competition. Rivalry with Apple, then with Kaypro Corporation,† and later with International Business Machines Corporation (IBM)† and its personal computer (PC) led to the decline of the company. Furthermore, other firms quickly recognized the demand for small computers, which eventually caused some 150 competitors to rival Osborne in the microcomputer market, often with better products than Osborne's. In September 1983 it declared bankruptcy and was reorganized; but its importance to data processing had come to an end. Like many of its counterparts in Silicon Valley, its one idea and one product had given it a moment of importance before it succumbed to management problems, newer and better technologies, and competition. While it functioned, it took computing at the personal level away from the hobbyists and hackers and placed it in the hands of a much broader community.

For further information, see: Paul Freiberger and Michael Swaine, *Fire in the Valley: The Making of the Personal Computer* (Berkeley, Calif.: Osborne/McGraw-Hill, 1984); Robert Levering et al., *The Computer Entrepreneurs: Who's Making It Big and How in America's Upstart Industry* (New York: New American Library, 1984); Adam Osborne and John Dvorak, *Hypergrowth: The Rise and Fall of Osborne Computer Corporation* (Berkeley, Calif.: Idthekkethan Publishing Co., 1984).

P

PHILCO. Philco was an important American manufacturer of consumer electronics and military hardware in the years following World War II. It also marketed computers during the 1950s before its acquisition in December 1961 by Ford Motor Company. Philco had revenues of $366 million in 1952. During 1955–1956 it constructed a computer for the U.S. Air Force, the transistorized C–1000, which was light enough to carry on an airplane and was a general-purpose device. This machine was followed by the C–1100 and the C–1102. In the same period Philco built a computer for the National Security Agency called the TRANSAC S–2000 which was replicated in civilian models. The first of these was the 210; others appeared in 1960 and 1961. Philco sold these devices primarily to the Atomic Energy Commission, General Electric,† Chrysler, United Aircraft, the California Department of Motor Vehicles, Ampex, and the University of Wyoming.

Philco's machine was likened to such powerful devices of the period as STRETCH* and LARC.* It was a general-purpose stored program, digital computer* that could be applied to scientific or commercial work and thus represented Philco's primary offering to the data processing industry. It gave the company considerable experience with computers, enough, for example, to allow it to bid against International Business Machines Corporation (IBM)† and Burroughs† on a contract in 1962 for NASA's data processing requirements to support the Gemini Program.

Although most of Philco's business was aimed at military projects, its presence in the market was not insignificant. Its revenues from data processing products in 1955 were $19.8 million and in 1963 had grown to $73.9 million. Its 1963 revenues exceeded those of some of the industry's giants: Burroughs, GE, Honeywell,† and the National Cash Register Company (NCR).†

Ford Motor acquired Philco as a means of establishing a foothold in the space and military markets. Beginning in the 1960s, Ford Motor provided more high-technology products to defense customers and to NASA because of Philco. The company recognized that Philco could not simultaneously participate as a full-fledged member of the data processing industry and be successful. Hence, Ford pulled Philco out of the industry to focus on space and defense. Like many other companies in that period, Philco had considerable potential to become a major force in the industry. Unlike many others, however, it took effective steps to establish itself well in the industry during the 1950s.

For further information, see: Robert Sobel, *IBM: Colossus in Transition* (New York: Times Books, 1981).

POWERS ACCOUNTING MACHINE CORPORATION. This company was Herman Hollerith's** primary competitor for card punch equipment from 1909 onward, and after 1911, when Hollerith's operations became the C-T-R Company (which in turn became the International Business Machines Corporation [IBM]† in the 1920s) it was the rival of the future giant computer company. IBM's arch rival during the 1920s was Powers' organization, which frequently had better products at competitive prices. The Powers Accounting Machine Corporation was formed by an ex-employee of the U.S. Census Bureau who, just before the Census of 1910, became convinced that he could market card punch equipment superior to Hollerith's and sell it to the Census Bureau for less. Encouraged by his improvements in card punch equipment for the Bureau and armed with a contract from that agency to supply equipment and cards, James Powers quickly became a significant competitor for Hollerith who, like Powers, had originally been encouraged to develop card punch technology by the Census Bureau.

The founder of this new company, James Powers (1871–1915), was born on February 12, 1871 in Russia where he graduated from the Technical School of Odessa after a solid grounding in mechanics. At age eighteen he moved to the United States where he worked in New York as a machinist for a variety of companies. These included the Carrin Machine Company, Western Electric and Bergman's Electrical Works. He expanded his experiences in precision machining methods and helped these companies develop a wide variety of equipment such as a letter-registering device, cash registers, adding machines and typewriters. He was a self-styled inventor who in addition to working on cash registers, adding machines, and typewriters, acquired patents for a toothpick-making machine, a breadbox made from glass, and photographic equipment. In 1907 he joined the Census Bureau to help improve on existing Hollerith equipment in anticipation of the next census. Bureau management was especially concerned about improving the speed and quality of its sorters, and that became Powers' first major project. While there he developed a new punch that looked like an oversized typewriter. He improved the efficiency of other card punches, sorters, and readers as well.

The Bureau quickly recognized and encouraged his talents, allotting him additional personnel and budget to expand development work. By early 1909 he had a new punch installed at the Bureau that was fully operational and that could process twice as many cards as Hollerith's older devices. His innovative technology had the potential of speeding up the Bureau's work for the forthcoming Census of 1910 while driving down personnel costs. Powers saw an opportunity to make money with his devices, and at the same time the Bureau perceived a good opportunity of breaking Hollerith's monopoly which it considered overpriced. The Bureau encouraged Powers to manufacture the new devices. Powers obtained patents on some of his innovations, despite a lawsuit initiated by the now embittered Hollerith. By the time World War I began, Hollerith's original patents expired, making it easier for others to enter the growing market for card punch equipment and supplies.

Just prior to the 1910 Census, Powers left the government to work for a newly formed company, Sloan & Chace. He had a contract with the Bureau to build devices he had designed. Quickly, the inventor next formed the Powers Company to compete directly with Hollerith in 1911. Powers continued to refine his products, particularly for sorters and tabulators. He began by relying completely on improved mechanical devices, and his company introduced electrical machines during the 1920s. He sold products to Hollerith's old customers both within the U.S. government and in industry just prior to World War I. During the war his orders increased, especially from the government. His company was unable to satisfy all of these orders, giving Thomas J. Watson's** company, the C-T-R (part of which consisted of Hollerith's company), an opportunity to sell products which sometimes cost more and were less efficient. During the 1920s technological innovations from Powers and IBM kept the competitive pressure on, creating a heated rivalry that IBM continued with Remington Rand† during the 1930s after Remington Rand had bought out Powers.

The 1920s witnessed the widespread use of card-tabulating machines throughout American industry and, to a lesser extent, in Europe. Part of this expansion in demand stemmed from technical innovations that made the equipment easier to use and less expensive. Powers contributed significantly to this situation. In 1924–1925 the Powers Company, for example, introduced its first alphanumeric punch and tabulator and a new method for doubling the amount of data that could be stored on a card.* IBM countered with its own products. In 1927 Remington Rand entered the card-tabulating business through the purchase of the Powers Accounting Company, taking advantage of a well-established company with a highly respected product line.

For further information, see: Geoffrey D. Austrian, *Herman Hollerith: Forgotten Giant of Information Processing* (New York: Columbia University Press, 1982); L. Couffignal, "Calculating Machines: Their Principles and Evolution," in Brian Randell, ed., *The Origins of Digital Computers: Selected Papers* (New York: Springer-Verlag, 1982): 145–154; L. L. Locke, "The History of Modern Calculating Machines: An American Contribution," *American Mathematics Monthly* 31 (1924): 422–429; J. H. Mc-

Carthy, *The American Digest of Business Machines: A Compendium of Makes and Models with Specifications and Principles of Operation Described, and Including Used Machine Valuations* (Chicago: American Exchange Service, 1924).

PRIME COMPUTER, INC. This company was one of the more important minicomputer manufacturers of the 1970s. It was formed in 1971, and it introduced its first product, the Prime 200, in 1972. The company experienced significant growth during the decade, and in 1979 it closed the year with revenues approaching $153 million. By the end of March 1980 it had shipped some 3,500 of its minicomputers worldwide. One of the company's more interesting financial characteristics was its investment growth. In 1974 its assets were valued at $7.5 million, which by the end of 1979 had grown to $142.7 million. With its good management, good products, relevant products, excellent timing, and little competition, Prime grew rapidly.

Additional products kept coming. During 1977 its major generators of revenue were the Prime 400 and Prime 500 processors which could hold up to 8 million bytes of main memory and 2.4 billion on disk—huge capacities even by the large computer standards of the day. By then the company also sold a variety of peripheral equipment to use with its computers. Its devices supported programs written in FORTRAN,* COBOL,* BASIC,* and RPG II (a favorite report-generating package for small business applications). Customers included small interactive business users, those interested in high-performance time-sharing systems, and engineering and scientific users. By the end of the decade, Prime introduced a networking package called PRIMENET which allowed its computers to communicate with each other, with those of other vendors, and through terminals over telephone lines.

The company was an early entrant into the minicomputer business and competed with International Business Machines Corporation (IBM),† particularly at the samll end of the processor business, and with other minicomputer manufacturers. IBM seemed to be a major target of concern. When the large manufacturer announced the 4300 series of midsized computers in January 1979, Prime reacted by bringing out compatible large processors (large by mini standards) called the Prime 450, 550, 650, and 750, all to compete with the 4331 and 4341 computers.

Prime continued to maintain a strong position within the minicomputer market in the 1980s, despite changes in management and in market tastes. In 1984, for example, it was the eighth largest minicomputer company as measured by annual sales. That year revenues reached $642.8 million, reflecting a 24.4-percent growth over the previous year on sales of data processing products. The growth in sales in the 1970s and early 1980s came largely from marketing to scientific and engineering users. Yet Prime moved into office markets in 1978 as well. Office products accounted for 20 percent of its total business volume by the end of 1981 and for 40 percent by the end of 1984.

The company's initial product offering, made in October 1972, rested on leading-edge technology, in addition to meeting the demands of a growing, if yet small, minicomputer market. Its earliest products were the first to use MOS memory, a common feature of larger computers. Subsequently, it claimed other minicomputer ''firsts,'' such as being the first to offer virtual memory minicomputers and the first to use 32-bit architecture in that arena. Despite these impressive achievements, earnings dipped in the early 1980s owing to an aging product line, or so at least industry pundits thought. During 1984 and 1985 Prime introduced new products, once again proving that failure to introduce new technologies and products rapidly could harm or destroy a company but, conversely, also offered opportunities for financial health.

For further information, see: Pamela Archbold and John Verity, ''A Global Industry: The Datamation 100,'' *Datamation*, June 1, 1985, 19, passim to p. 182; Franklin M. Fisher et al., *IBM and the U.S. Data Processing Industry: An Economic History* (New York: Praeger Publishers, 1983).

R

RADIO CORPORATION OF AMERICA (RCA). This long-established electronics firm entered the data processing industry full-force during the 1950s and left it at the end of the 1960s. During the period of its greatest impact on the industry (1960s), it was often called one of the Seven Dwarfs, a reference to some of the major computer vendors in competition with International Business Machines Corporation (IBM).† When RCA first introduced computers into the American economy, many industry watchers considered this company, rather than IBM, as a prime contender for the position of leading computer manufacturer. A number of factors, however, forced the company to exit from the computer market at the end of the 1960s. Although its role within the industry would appear brief, its technical work spanned many decades and its influence on events within the field of computers was often intense and important. Moreover, its history reflects many of the concerns and patterns of behavior evident in other computer manufacturing companies.

RCA was founded in 1919 after the radio became a commercial and technical success. General Electric (GE),† which was then negotiating with the Marconi Wireless Company of America, a British firm, was interested in the right to use British technology. The move was blocked by the U.S. government, which was reluctant to see patents for such technology possibly move out of the country. The opposition of the government to such negotiations led in 1920 to the creation of RCA for the purpose of handling radio business by a GE executive, Owen D. Young (1874–1962). In this deal RCA acquired the rights and business interests of the American Marconi firm and a complicated set of cross-company rights involving such other firms as Westinghouse and GE necessary to launch commercial radio in the United States. Throughout the next decade RCA made and sold radios and radio-related parts and equipment. In the process it built up a nucleus of electronic skills which in future years would make possible the company's quick move into computerized technologies. It allowed the introduction of television and computer-related products. An event only partially related to computers, but nonetheless significant for RCA, was the establishment of the National Broadcasting Company (NBC) by RCA's general manager in 1926, David Sarnoff (1891–1971). In the decades to come NBC would cause the company to focus much attention and resources on entertainment, radio, and television often in competition for the same managerial talents and investment dollars that might otherwise have gone toward computer-related research.

During the 1920s and 1930s, scientists at RCA did considerable research on broadcasting and electronics. During the 1930s in particular, RCA supported an active research program developing television transmission and reception which was interrupted by World War II but picked up almost immediately after 1945. By the end of the war, RCA was a major electronics firm. Like many other such companies, it had been associated with the government on various technical projects. RCA's scientists, for example, developed a firm control computer, some of whose functions were later adopted by the Moore School of Electronic Engineering† for use in the ENIAC.* Dr. Jan A. Rajchman** of RCA played an early and important role in the development of magnetic core technology later used in computers to store information and later worked at the Princeton Institute for Advanced Study on computers. While at RCA he and other scientists also worked on the development of a storage tube that was later used in the Rand JOHNNIAC* computer. Later he and others worked on various vacuum tubes. After he left RCA, others carried on work that eventually led to a variety of transistors.

In 1947 RCA built its first major computer for the U.S. Navy. Although it had previous experience with digital computers,* this project represented the first major step in the creation of an analog device. Almost simultaneously, RCA explored the commercial possibilities of building a digital computer. It became an early contractor for the SAGE* defense project during the 1950s. RCA's experience making vacuum tubes for radios (also used in the early days of computers) insured that it would be consulted on the ENIAC project. Work done in the late 1940s with transistors sped up the transfer of data within computers while increasing their reliability dramatically. Research continued during the early 1950s on core memories, particularly to satisfy government contracts.

RCA was thus poised to enter the computer field by the early 1950s. In addition to its scientific capabilities, it was a large firm with many resources, including plants used for the production of scientific (radio) devices. Revenues in 1952 approached $694 million. Using its resources, RCA decided to develop a commercial computer, and after several other companies had started to ship computers to commercial customers, RCA in 1956 announced its BIZMAC.* Univac and IBM already had products in the market. The BIZMAC, a large processor by the standards of the day, sold for $4 million. It was designed to handle business applications such as accounting and inventory control. As with many of the early computer projects, this one grew out of research connected with a U.S. government contract, specifically a contract to build a computer for the Army to use in inventory control. Technically, the product ran slower and less efficiently than other computers available in the market. As a result of this technological inferiority, only six copies were sold.

In 1958 work began on the 501, and at the end of the year RCA announced the product. It, too, was a general-purpose computer, and yet it was based on transistor technology. By the end of the 1950s RCA had delivered three copies to customers. The failure of this product was not so much due to its technological features as to the poor quality of its peripheral equipment. The card readers and punches, for example, were too slow and unreliable by the standards of the day. The system's printer was considered of poor quality by prospective customers, and it broke down too often. Random access storage for the system, called the RACE, also proved too unreliable to be competitive. Thus, the Computer Division of RCA closed out the 1950s without having made a significant impact in the data processing industry, despite the fact that the company had been well positioned to do so in the late 1940s.

The chief executive officer at RCA in the late 1950s was John Burns, a man of considerable experience. Just prior to joining RCA in March 1957, he had served as consultant to IBM where he had been privy to all of its future development plans for computers and for marketing them. After he joined RCA, competition between the two companies took on an almost personal note between the two executive officers, with IBM's Thomas Watson, Jr.,** annoyed that so many company secrets could potentially help a rival. Indeed, Burns concluded that building computers at RCA was a critical opportunity not to be missed. However, this intent was heavily compromised throughout the 1960s by internal competition within RCA for research funds between those advocating work on computers and others pushing for the creation of color television. This rivalry resulted in insufficient funds being channeled into data processing at a time when companies such as IBM were prepared to gamble everything on the future of computers. RCA's computer developments during the 1960s constitute a sad tale of frustrations, missed opportunities, and tardiness in product introductions.

In 1960 RCA announced the 601 as a large general-purpose machine and the 301 for a medium-sized computer market. The 601 cost more to manufacture than it could be sold for, and it never had all the functions which RCA promised at its announcement. Furthermore, it was plagued with technical difficulties. For example, cabling the computer became too complicated and bulky. In 1962 it was dropped from the product line; only four were ever delivered. The 601 had been an expensive failure that also hurt RCA's credibility within the industry. Customers with 301 computers who wanted to upgrade could not do so once the 601s were withdrawn from marketing, giving them yet another reason not to do business with RCA. Work on peripheral equipment was cut back at this same time, particularly in 1961, in order to focus more effort and funding on the development of mainframe computer technology. In a switch in strategy, the following year RCA opted once again to build its own peripheral devices rather than buy them from other vendors in an attempt to increase the reliability of such equipment to levels it sought. Furthermore, RCA wanted to control the technical functions incorporated in such equipment, particularly in data storage medium that was becoming increasingly competitive.

A milestone in the history of computers was reached in April 1964, when IBM announced the S/360* family of computers. For the first time a vendor made available various sizes of computers which were all compatible, that is, programs operating on one model could run on another without expensive and time-consuming conversions. This announcement also offered a variety of major technical improvements in the hardware then available and in previously used programs. The new products forced other computer vendors to react with their own families of computers or with major price reductions of existing product lines. RCA reacted by cutting prices between 20 and 35 percent. Next, the company announced the Spectra 70 family of computers which were compatible with IBM's 360 processors. The intent was to go after IBM's market. The pricing for these machines was at times lower than IBM's. RCA's announcements of new computers represented the data processing industry's first major competitive response to the S/360.

Until 1964 RCA had experienced considerable difficulty in gaining customer acceptance of its 601. Its only other viable computer product was the 3301 which had been introduced in August 1963. This product did poorly owing, first, to the lack of technologically current and reliable peripherals and second, to the announcement of the S/360 which in turn produced the announcement of Spectra. Customers elected to bypass the 3301 system and either switch to other vendors (such as IBM which almost doubled its revenues before the end of the 1960s) or to Spectra. Thus, Spectra killed off the old products at RCA while ushering in a new era for RCA's computer division.

By the end of its product life, Spectra comprised eight models (70/15, 70/25, 70/35, 70/45, 70/46, 70/55, 70/60, and 70/61). The last model was announced in 1969. Deliveries of the earliest machines did not begin until mid–1965. Spectra's compatibility with IBM's machines indicated that RCA wanted to go after the larger firm's rapidly growing install base. Prices were structured to be between 15 and 20 percent below comparably sized IBM devices. The computers were sold for both commercial and scientific uses.

Despite its carefully designed marketing strategy, RCA experienced difficulties with Spectra. First, the equipment required more preventive maintenance than IBM's machines. The same applied to both disk drives and computers. During 1968 RCA's officials reported that 75 percent of the marketing force's time had to be spent coping with problems in existing Spectra installations instead of in selling additional copies of these machines. Problems with hardware downtime and the computer's oversensitivity to electricity and air conditioning which caused additional failures drove maintenance costs up sharply. Software problems in the late 1960s also plagued the company as it attempted to expand its offerings into time-sharing products. Throughout the life of the Spectra family, RCA's memory stacks never functioned as desired.

RCA's computer sales between 1960 and 1965 rose from $14.6 million to over $100 million. Orders for computers in 1965 were 90 percent over 1964's total. There was a generally increased demand for all computers across the entire economy. Expansion could also be attributed to aggressive pricing and to an increase in RCA's computer-related sales force of about 45 percent in 1965 and an additional 35 percent in 1967. Costs rose quickly, too, keeping computer sales unprofitable. Revenues remained high, although not profitable; from 1965 to 1970 they went from $89 million to almost $211 million.

With funds still pouring into computer development and sales hampered by technical difficulties, despite increasing volumes of revenues in the late 1960s, the company underwent internal changes and by the early 1970s RCA dropped out of the computer business. In January 1968 Robert Sarnoff became the chief executive officer. Other executives were brought into the highest echelons of the company, including a new marketing vice-president and president. With new management, the company's diversification accelerated. Its products and services changed as it expanded its line from television and electronics to include home appliances, communications, broadcasting, food, household goods, real estate, and car rentals. Hertz (a cash cow for the company), Random House (publishing), and Coronet (home furnishings), for example, were all part of RCA during the late 1960s and early 1970s. As an example of the contribution of data processing sales to this new RCA, in 1971, revenues from data processing products accounted for only 10 percent of RCA's revenues of $270 million.

Internally, then, competition existed for attention and resources that might otherwise have gone toward computers. In those years performance was measured primarily by market share and speed of market penetration. These measurements often became more significant for the careers of executives than the more traditional profits earned and expenses controlled. Thus, executives in the Computer Division were motivated to expand their business with less regard to cost than might otherwise have been the case. In order to expand business, in 1968 they decided to begin work on a follow-on product line to Spectra. Investments in research, however, were dispersed across two development projects, each of which was intended to generate the next product line and at a time when other divisions within RCA were demanding additional resources. Of the two major research projects, the more important was the New Technology System (NTS). This product promised major enhancements over developments of the past and made possible management's decision to market a broad line of products that would totally replace Spectra and its peripherals. These new products were also designed to compete with what RCA's management expected would be IBM's follow-on product to the S/360 which became known as the S/370.*

RCA announced new products in September 1970 in reaction to IBM's announced S/370 of three months earlier. The new devices were called the RCA 2, 3, 6, and 7 and were medium-sized computers. Briefly put, they were more powerful than Spectra machines and less expensive. As officials within RCA later admitted, these new machines were repackaged Spectra 70 computers with some new memory technology added. The RCA series proved to be a major product disaster. First, this announcement caused a sharp drop in the sales of the more expensive Spectra series rather than the displacement of IBM's installed S/360s. Second, the new machines continued to be plagued by technological difficulties.

RCA's customers could save about 15 percent of their computer costs by swapping their Spectras for the RCA series. They would also be installing devices that were larger and supposedly more efficient. This thinking led to a decline in RCA's rental income, while causing the company's inventory of returned Spectras to rise and at a time when these machines were overvalued on the books—all costly to RCA. Seventy percent of all Spectras returned were due to swapping of computers with the RCA series. Only 18 percent were due to competitive losses. Thus, the company's greatest enemy was not IBM but itself. Inventory now had to be written down in the books faster than originally anticipated, causing RCA's profits to decline.

Technological difficulties abounded and were serious. Peripherals were not state-of-the-art and, when compared to IBM's then available, were two to three years behind, particularly with disk storage. RCA's peripherals, which in many cases had been introduced to compete head on with IBM's, could not compete, especially with IBM's 3330s which had been introduced in June 1970 and would become the industry's standard for over a half decade. Similar problems existed with tape drives. Delays in delivering tape and disk drives along with more advanced software threatened even further loss of sales. Between 1971 and the end of 1973 the company estimated that technical problems probably cost it $3.5 million in revenues.

Thus, RCA's Computer System Division entered the 1970s with troubles that it never managed to overcome. Losses were forecast to rise to nearly $35 million in 1971 if corrections were not made within this division. Top management realized the causes. First, revenues were declining owing to the lesser cost charged per unit shipped of the RCA series. Second, fewer Spectras were sold. Revenues were slipping from a high of $230 million to $186 million. Already mentioned was a third problem, the high costs of manufacturing and marketing the new machines. RCA's manufacturing expense as expressed as a percentage of revenue neared 42 in contrast with IBM's 14 to 15 percent. Sperry Rand's† percentage of manufacturing expense to revenue hovered at 24 percent and Burroughs† at near 21 percent. Some industry observers noted that RCA had poor internal information systems with which to track the maintenance history of its machines, to do customer surveys, and so on, to guide management in understanding their business, and for making better decisions. When poor information support systems were coupled with inadequate financial controls, top management above the division level realized that computer sales were hurting RCA too severely.

As this review was taking place, technical and manufacturing problems compounded, causing late product shipments. Competition from IBM increased primarily because the S/370 family of computers was priced lower than RCA had anticipated and these machines enjoyed fewer technical difficulties. They were also installed closer to their original target dates than RCA's products, thereby generating revenues for IBM on a more timely basis.

Thus, by mid–1971 RCA's computer business had reached a crisis point. Rumors now spread that RCA would get out of the computer business. Top executives denied the rumors—yet by September they decided to do just that. The decision was sudden, was made by the board of directors, and the computer division was never consulted. That division's employees were caught by surprise, as were competitors and customers. But the decision was understandable. Management estimated that losses would soon have run over $100 million. Board members began to realize that to remain viable in the industry, RCA would have to invest another $700 million in data processing between 1971 and 1976. This investment would have come at a time when competition within RCA for funding other business sectors had become important. Pressure had already mounted to fund additional research and development of new television products. Furthermore, computer sales accounted for only about 6 percent of RCA's business, even though that share could have been expected to grow with the additional investments to as much as 15 percent at best. The case for staying in the computer business did not look attractive enough, particularly when measured against RCA's previous performance in that arena.

RCA sold its computer division to Sperry Rand. RCA reported its losses on computers before taxes between 1958 and 1971 had been $241 million. And, in that final year, it set up a fund of $250 million after taxes to cover anticipated losses resulting from the sale of the division. In November RCA announced that Univac would obtain the customer base, cutting off much speculation about who would maintain existing machines and service RCA's accounts. The issue of support was critical for those who installed equipment with a history of technical problems. Customers for years afterwards were rankled over RCA's inability to resolve the maintenance issue in the fall of 1971. When developing their companywide and account plans, many computer vendors afterwards took careful note of that concern.

The announcements were the culmination of many months of tension within the entire industry. As a result both of rumors that RCA was pulling out and of real problems within the company, its computer sales had declined during the summer of 1971. Customers had become fearful of being "stuck" with a computer that might not be serviced and whose value was declining sharply. The leading plug-compatible manufacturer was in trouble, and RCA's customers knew it. Honeywell and IBM in particular noticed increases in sales as customers shifted away from RCA. RCA's leases were being canceled, thereby increasing pressure on top management to make their decision in September.

The data processing industry learned a number of lessons from RCA's performance. Building IBM plug-compatible machines was possible, but doing it efficiently was not easy. Customers now paid more attention to the economic well-being of their computer vendors when making purchasing decisions. For the manufacturers of data processing equipment, it was a lesson on how significant the investment had to be if cost-effective products were to be sold profitably.

In retrospect, RCA's failures were due to a number of factors. Management was poor and, more importantly, had little experience with data processing, especially at corporate headquarters. Too much diversification within RCA made it difficult, if not impossible, for the company to make the kind of commitment to computers that it would take to be successful. In short, RCA did not have the singlemindedness that was so evident at IBM. Finally, the company did not have enough resources to fund all the necessary projects within RCA. It could make a bigger impact in defense electronics and television, where it had strength, at lower costs.

Throughout the mid–1970s, RCA did not completely leave the data processing world. Research progressed on chip* technology, and RCA sold airline reservation terminals and built specialized processors for defense contracts. But none of these activities was sufficiently dramatic or large to influence the data processing industry as it had with its earlier work with plug-compatible computers. Meanwhile, UNIVACs and other computers slowly replaced Spectras and RCA series installed throughout RCA's old customer base. By 1975, 77 percent of RCA's old computer customers still had RCA equipment. By May of that year it was down only 1 percent, but the decline became more rapid as machines were fully depreciated, leases ran out, and newer technologies became available.

Missed opportunities, poor management, little appreciation for the data processing industry, and inability to introduce new technology at better prices in a more timely fashion shook RCA's confidence throughout the 1970s. Only by the end of 1983 had the company reestablished a clear and simple direction. The previous decade had witnessed changing identity and missions. By 1984 RCA was returning to its strengths: electronics, communications, and entertainment. Sales over the previous decade had only doubled, from $4.2 billion (1973) to $9 billion (1983), with a long-term profit margin of 2 percent. This performance lagged behind that of other large high-tech companies of the period.

In the mid–1980s RCA had considerable data processing technical expertise which it applied to communications, defense work, space, and television. At that time, RCA showed no signs of wanting to reenter the data processing field on a large scale as it had during the 1950s and 1960s.

For further information, see: Franklin M. Fisher et al., *IBM and the U.S. Data Processing Industry: An Economic History* (New York: Praeger Publishers, 1983); Katharine D. Fishman, *The Computer Establishment* (New York: Harper & Row, 1981); Herman H. Goldstine, *The Computer from Pascal to von Neumann* (Princeton, N.J.: Princeton University Press, 1972); J. Rajchman, "Early Research on Computers at RCA," in N. Metropolis et al. eds., *A History of Computing in the Twentieth Century* (New York: Academic Press, 1980): 465–469 and his *RCA Computer Research: Some History and a Review of Current Work* (Princeton, N.J.: David Sarnoff Research Center, RCA Laboratories, 1963); Robert Sobel, *IBM: Colossus in Transition* (New York: Times Books, 1981) and *RCA* (New York: Stein and Day, 1986).

RADIO SHACK. See TANDY CORPORATION

RAYTHEON COMPANY. Raytheon was one of the first companies to manufacture computers in the United States, beginning in the late 1940s. It dropped out of commercial data processing in 1984 when it abandoned direct sales to commercial customers. Like many of the early companies in the data processing industry, it first established itself as a supplier of electronics to the U.S. government and then expanded into data processing technology in general.

Until the start of World War II, Raytheon was a small electronics supplier. During the war it expanded, building radar and other electronics for use by the military. As a result of that experience, it obtained contacts willing to buy computers in the late 1940s. Second, it built up considerable in-house expertise in leading-edge electronics. Its first foray into data processing came in 1947 when it began work on the Raytheon Digital Automatic Computer (RAYDAC)* which it was constructing first for the U.S. Bureau of Standards and, subsequently, for the Office of Naval Research (ONR).† Work on the machine was not completed until 1951, by which time it was obsolete. Raytheon built other computers for classified users, usually in the national security community of the U.S. government. As a result of these projects, this major supplier of government electronics acquired experience with the fundamental building blocks of computers: transistors, triodes, Klystron tubes, and later chips.* In the 1940s and early 1950s all of its data processing products went to government agencies. Various U.S. agencies considered Raytheon's skills in the early 1950s equal to Univac's. In 1952, for example, Raytheon became one of a few companies selected to work with the Lincoln Laboratory† at MIT on the design of the SAGE* computer system.

Like other companies that started out with the government as its customer, Raytheon sought to broaden its base by offering products to commercial users of computers. In 1953 the company began exploring that market and initiated work on a general-purpose commercial digital computer* which it called the RAYCOM, a commercial version of RAYDAC that was never completed. The company elected to change directions, choosing to stay out of commercial data processing and going back to supply only the government.

Yet in 1955, taking advantage of its broad knowledge of data processing, Raytheon and the Minneapolis-Honeywell Regulator Company (soon after called Honeywell†) agreed to establish a joint venture called Datamatic Corporation. Datamatic was founded to develop large computers for commercial customers relying on the technology in RAYCOM. For Honeywell it offered entrance into the computer industry while Raytheon sought access to commercial customers. Raytheon owned 40 percent of Datamatic and had all the expertise in computer-related technology. Honeywell retained general management of Datamatic and responsibility to raise capital and sell its products. Datamatic's first commercially available system was the Datamatic–1000 followed later by the Honeywell 800 and 400 computers. It also played an important role in the development of the Honeywell 200 system.

Although Raytheon missed an opportunity in the early days of computing to become a major force in the industry, Honeywell did not. Honeywell's management recognized that Raytheon's technical skills could transform it into a major computer vendor fast. In 1955, the year the Datamatic project began, Honeywell was already well positioned. It had sales of $244 million, with net income of $40 million before taxes, and was recognized as an important vendor. Its total assets amounted to $164 million. In that same year Raytheon clocked in sales of $182 million with pretax earnings of $9 million. It controlled assets worth $82 million. Put another way, the total earnings and assets of these two companies were similar to those of Sperry Rand† or International Business Machines Corporation (IBM).†

Datamatic prospered. Its first product, the D–1000, was a large first-generation computer that relied on vacuum tube technology. It was first shipped in the last quarter of 1957 and had a price of nearly $2 million. Honeywell supplied its peripherals or acquired them from other vendors, including card equipment from IBM. Honeywell sold nine of these systems to commercial customers. In 1957 Honeywell sought to bolster its position with computers and so bought out Raytheon's 40 percent share of Datamatic for nearly $4 million. Raytheon agreed to this sale because it was not prepared to make the significant investments in Datamatic it felt would be required to produce additional computing products. At the time, Lockheed, also interested in buying the firm, had concluded that, unless Datamatic made significant technological improvements soon, its products would become dated and therefore not competitive. Datamatic's subsequent history is the story of Honeywell's data processing activities of the late 1950s down to the present.

Raytheon's focus turned back to specialized equipment for the government. During the 1960s and 1970s it continued to be a major provider of electronics to the military. Much of these electronics involved the use of data processing technology. The company also manufactured components for other computer vendors producing, for example, chips.* During the 1960s it introduced a line of cathode ray terminals (CRTs) that were largely used in airline reservation systems by the 1970s. These terminals accounted for one-sixth of Raytheon's Data Systems' annual sales in 1983; 1,500 people were employed just to maintain this particular set of products.

During the 1970s Raytheon was back in the computer field on a limited basis with commercial products, but on May 18, 1984, it announced its second withdrawal from that part of the industry, citing competition from more established vendors as the reason. By then the company's total annual sales had reached $1.5 billion. Its data processing operations, all gathered together in Data

Systems, Inc., sold various items, but, by the early 1980s primarily terminals. In 1983 it had generated sales of $307 million, amounting to 5 percent of all sales at Raytheon. That same year Data Systems lost $24.3 million and, before it was announced that it would reduce operations beginning in May 1984, had lost an additional $6.2 million during the first months of the new year.

Other major portions of Raytheon included Beech Aircraft Company (acquired in 1980). Its United Engineers and Constructors, Inc., did not post losses of the magnitude of Beech Aircraft's in 1984, but it did face lawsuits associated with cost overruns in the construction of nuclear power plants. Meanwhile, sales of military electronics remained solid. But diversity meant that the small data processing operations at Raytheon would not receive adequate attention. And without significant commitments to data processing, the company could not hope to compete effectively with companies such as IBM and Honeywell which had committed their total resources to data processing. Thus, Raytheon is of interest for its early role in data processing and as a source of speculation about whether it could have been a major factor in the data processing industry.

For further information, see: Franklin M. Fisher et al., *IBM and the U.S. Data Processing Industry: An Economic History* (New York: Praeger Publishers, 1983); Katharine Davis Fishman, *The Computer Establishment* (New York: Harper & Row, 1981).

RCA. See RADIO CORPORATION OF AMERICA

REMINGTON RAND CORPORATION. Remington Rand was a major force in the office equipment market during the 1920s and, in the years following World War II, merged with Sperry Corporation to form Sperry Rand Corporation.† Remington Rand was the source for the UNIVAC* series of computers which helped create a commercial computer market in the early 1950s. This company was a collection of small typewriter and other supply companies for office equipment. These little firms had played various roles in the development of data processing as an industry for the office and government. The history of Remington Rand thus reflects many of the patterns evident in the office machine arena from the turn of the century to the present. Its current form, the Unisys Corporation, remains a major force in the data processing industry today.

Remington Rand was formally incorporated on January 25, 1927, in Delaware when it acquired the stocks of the Remington Typewriter Company, Rand Kardex Bureau, Inc., the Dalton Adding Machine Company, the Safe Cabinet Company, and the Powers Accounting Machine Corporation.† Although it has a complicated history and parentage, 1927 represents the company's true birth year. It was the product of James Henry Rand,** who had successfully managed a number of business machine ventures. Together with Irving Fisher, economics professor of Yale and in the 1920s a popular analyst of Wall Street's activities, as well as other investors, he established a series of smaller companies which formed a corporation that could challenge companies such as National Cash Register

(NCR)† and Burroughs.† The merger might also have been initially motivated as a way to profit from stock transfers—a charge also made about the C-T-R arrangement just before World War I that would eventually become the International Business Machines Corporation (IBM).† The product lines of each of these smaller entities complemented each other—certainly more so than in the case of IBM's predecessors—and so the same salesmen could sell typewriters, office machinery, furniture, and office supplies effectively. The Powers Accounting Machine Corporation had been Herman Hollerith's** old rival for card punch products at the turn of the century. Kardex had been (and would continue to be until the early 1960s) a popular card filing system widely used in industry, business, and government agencies. The system was personally developed by James Rand around 1915. The Dalton Adding Machine Company was typical of many competitors who rivaled the much larger Burroughs and NCR in the early years of the 1900s.

The Remington Typewriter Company, which gave part of its name to the new firm, had been incorporated on May 25, 1909, in New York, under the name Union Typewriter Company and in March 1912 was renamed the Remington Typewriter Company. This firm had, in turn, acquired other office machine companies in existence in the late 1800s, including the Monarch Typewriter Company, the Yost Writing Machine Company, the Smith Premier Typewriter Company, and the Densmore Typewriter Company. In 1924 it bought the Remington Noiseless Typewriter Corporation which had succeeded the Noiseless Typewriter Company, Inc. Sales dropped from $3.3 million in 1919 to $1.5 million in 1922. By the mid–1920s, however, sales had stabilized, and its assets had grown in value to $29.6 million, making it a candidate for a corporate takeover.

The creation of this company resulted from the consolidation of many smaller firms that dotted the American office machine and supply landscape in the late 1800s. Expressed in its most negative connotation, one could argue that reading a history of Remington Rand was like reading the obituary column of bygone companies. In the late 1800s industrialization was in full bloom, and the office was undergoing changes as radical as those it would experience in the 1980s. Both eras experienced a surge in demand for information, as well as systems and technologies to manage such data. Although labor costs were less of an issue in the 1800s than in the 1980s, automation to improve human productivity was an attractive feature of office equipment and information-handling technologies. Time and speed, not to mention accuracy and quantity, were crucial issues. Automation provided real opportunities then as it does today. In the last half of the nineteenth century, and particularly in the last quarter of the 1800s, offices acquired adding machines, typewriters, telephones, fountain pens, many of the standard forms in use today (e.g., preprinted invoices, payroll checks), electricity, and even furniture still in evidence today (swivel chairs, desks with multiple drawers, and file cabinets of the size in use today). That process of change continued at a more rapid rate in the early twentieth century, only to be slowed temporarily by the Great Depression of the 1930s.

That round of technological change created the demand for products, many of which were new, justifying the establishment of many office supply manufacturing firms. Their names and births reflect the growth in the demand for the paraphernalia of the "modern office": Remington Typewriter Company (1874), Library Bureau (1876), Rand Company (1876), Safe-Cabinet Company (1887), Baker-Vawter Company (1888), Dalton Adding Machine Company (1902), Kalamazoo Loose Leaf Binder Company (1904), Noiseless Typewriter Company (1904), Remington Accounting Machine Corporation (1907), Index Visible, Inc. (1911), Powers Accounting Machine Corporation (1911), Accounting and Tabulating Machine Corporation (1911), Lineatime Manufacturing Company (1914), and the American Kardex Company (1915)—all of which became part of Remington Rand in its early days. By 1932 this collection had over 11,000 employees with plants totaling 3.8 million square feet.

Low profit margins in a competitive industry made consolidations attractive. The economies of scale made possible were also real. The result in this case was a large company which, by the end of 1928, had sales of nearly $60 million. In contrast in that same year IBM had sales only about one-third of that, although their profits were nearly equal ($5.3 million for IBM versus Remington Rand's $6 million). The feeling at the time, particularly in the business press, was that Remington Rand could fully displace the giants of the office machine market, Burroughs and NCR, within a few years. In 1928 total revenues made it first among all business machine companies, followed by NCR (at $49 million) and Burroughs (at $32.1 million). Throughout the 1930s it battled for position but failed to gain the dominance predicted in the 1920s. By 1939 IBM's better management had catapulted that company into second place after Remington Rand. NCR was in third place with $37.1 million in sales and Burroughs in fourth with $32.5 million. By the end of World War II (during which Remington Rand had devoted considerable effort to the manufacture of ammunition), IBM emerged number one with sales of $141.7 million. Remington Rand was second ($132.6 million). IBM's earnings had also jumped dramatically over Remington's, just as did its net assets. It appeared that Remington Rand might be on a slide downhill for market share.

With such size and diversity of products, the company entered the late 1940s in a good position to take advantage of computers and to become a giant in that fledgling industry. It had a sales force and a broad set of loyal business customers, not to mention good penetration within the government, particularly the military. The company also had considerable expertise with electromechanical unit record card equipment. In the late 1940s it introduced new products, including an electric typewriter as did other firms. Company chairman Rand, still the innovator, was interested in computers and kept up with developments.

His strategy for entering the computer market was the same one which the company used at its formation: acquisition. Beginning in 1950, he acquired the Eckert-Mauchly Computer Corporation and, two years later, Engineering Research Associates†—two leaders in the new technology. Those acquisitions made Remington Rand the giant of the new computer manufacturing industry by 1952, a position it would relinquish to IBM by the mid–1950s as it had a decade earlier for office equipment.

Two engineers at the Moore School of Electrical Engineering† at the University of Pennsylvania, J. Presper Eckert, Jr.,** and John W. Mauchly,** had built the first working electronic digital computer* in the mid–1940s called the ENIAC.* They left academic life in 1946 to form their own company, with the purpose of manufacturing computers for sale to government agencies and corporations. Building on the work they had done on the ENIAC and the follow-on machines (EDVAC* and BINAC*), they began to design what eventually became known as the UNIVAC, working on the machine throughout the late 1940s. Their company was never properly capitalized, however, to continue research on this computer for so long. Plagued by cash flow problems, the two eventually sought out a buyer for their firm in order to continue work on the computer. The chairman of the Remington Rand Company had been aware of their work and made known his desire to buy their operation. By the time Rand did this, the UNIVAC, intended to be manufactured rather than be built as a one-of-a-kind device, was in its final stages of development. Within a year of acquisition, the first UNIVAC was delivered to and accepted by the U.S. Bureau of Census for a cost of approximately $1.1 million. In 1950 six were under contract, and Eckert and Mauchly estimated that they only needed an additional $1 million in capital to complete UNIVAC's development and manufacture.

Remington Rand entered into negotiations, despite industry claims that the market for such computers was minuscule, and concluded the purchase in 1950. Eckert and Mauchly sold 95 percent of their stock to Rand and in exchange were given annual salaries of $18,000 each (they had been paying themselves $15,000) and continued work on the computer as executives in the division they now made up in Remington Rand. Following the conclusion of this purchase in March 1950, the company funded the rest of their research. The UNIVAC became an instant success, so much so that by the mid–1950s the word *UNIVAC* was almost synonymous with the term *computer*, as Xerox† would be with photocopying in the 1960s. The Census Bureau's UNIVAC was extremely useful in tabulating 1950's data. The final 5 percent of stock outstanding for the Eckert-Mauchly Computer Corporation was purchased, and additional research funding was allocated to the UNIVAC project. Five more machines were built for the Census Bureau, and the forecast for demand rose to twelve devices. With the Korean War now in progress, it appeared that the military might need some UNIVACs, so the prospects for the new computer industry kept growing. In 1954 General Electric† became the first commercial enterprise to buy a UNIVAC, making it the first commercial customer to acquire a computer for business use. As a result of this sale, the future never looked brighter at Remington Rand. The company sold nearly forty copies of the UNIVAC before it brought out the UNIVAC II in 1957, proving that the company was in a rapid growth industry.

Stepping back to 1950/1951, Remington Rand was encouraged sufficiently by the UNIVAC project and the potential for such technology to search for other acquisitions. In 1952 it bought Engineering Research Associates (ERA) of St. Paul, Minnesota, a firm that had built a computer for the Georgia Institute of Technology. ERA was then preparing other versions of the machine for the military services. Much like the Eckert-Mauchly Computer Corporation, ERA was formed in 1946 by people with backgrounds in technical fields. ERA was run by William Norris** who after World War II continued on communications and computer projects for the military. To insure that his group of technicians would not disband after the war, the government encouraged Norris to form his own company with the implied recognition that various agencies would supply contracts to keep it going. The U.S. Navy was its primary customer with whom it had a contract to build the ATLAS I computer. A commercial version of the machine, named the 1101, was first shipped to a commercial customer in December 1950, nearly four months before the first UNIVAC went to the Census Bureau.

It also made the 1102, a general-purpose computer which sold for $575,000 in 1952. Only three copies of each system were sold. Yet development work continued. Even as early as 1949, the 1103 was being designed, and eventually it was announced in 1953. Remington Rand sold twenty of them. The acquisition of ERA and its 500 employees, many of whom were technically outstanding, made Remington Rand a major potential force in the computer industry. By the end of 1952 ERA alone had shipped nearly 80 percent of all the computers then in use within the United States.

Because of Remington Rand's preeminent position at the start of the 1950s, it made perfect sense for both Eckert and Mauchly, and later Norris at ERA, to join forces with the office machine giant to maximize their abilities to market commercial versions of their machines. Yet by the mid–1950s, IBM's rapid product development superseded theirs. Technical and marketing leadership, often a delicate balance in this industry, therefore, slipped from Remington Rand. Why did this company lose its formidable lead? Its resources were impressive, and it had a loyal customer set to begin selling to such devices.

A number of answers have been given to this question. First, James Rand, known as an autocrat and an "iron-willed manager," although he recognized some potential for computers (at least enough to buy these two companies), did not foresee the real demand that existed and therefore the commitment that would be required of his company. He thus failed to promote the right kind of marketing and development plans that together would have insured his company's preeminent position. Unlike IBM, which bet its entire future and committed all of its resources and management talent to computers, the subject received less than adequate attention and support within Remington Rand. Thus, new products came out too slowly. For example, the UNIVAC II appeared in 1957, too late and with too little new technology. IBM's products were technically advanced by then and were sold by a company that was better organized to market computers. It was also armed with a knowledgeable and motivated sales force.

Historians have concluded that Remington Rand failed, first, to develop marketing strategies or a knowledgeable sales force to sell computers. Thus, for instance, its salesmen continued to push typewriters and punched card equipment with little incentive to market computers. Second, it failed to educate customers on why they needed computers or how to use them—a mistake IBM avoided. Third, inadequate product development (for instance, later members of the UNIVAC series) meant that its arch rival, IBM, could take the technical lead in 1954/1955. Remington Rand records and legal testimony by its officials indicate, however, that it did have the resources to invest in computers. Fourth, employees disenchanted with their company's lack of full commitment left the firm and joined other new companies coming into the data processing industry. Norris, for example, left in 1957, and Mauchly in 1959. Fifth, the two computer firms acquired by Remington Rand were never fully integrated—a mistake evident again in the 1970s with its own product line and that of Radio Corporation of America (RCA).† Therefore, economies of scale, the synergism that could have existed with technology, and marketing were not there. Finally, weak management crippled the firm's ability either to commit itself to the new industry or to sustain and improve profitability. More attention, it seemed, was focused on acquiring companies (including the capability to manufacture electric shavers) than in managing a serious and complex operation in a complex new industry.

The lead passed to IBM, and, later, to others as well. Statistics indicated the speed with which this happened. In August 1955 Remington Rand had placed more UNIVACs than IBM its own machines (700 series). One year later, IBM had installed seventy-six computers, and Remington Rand forty-six. IBM had 193 on order, and its competitor 65.

The final chapter was written on the old Remington Rand firm in 1955 when it merged with the Sperry Corporation to form a new company called Sperry Rand. James Rand was named vice chairman, and General Douglas MacArthur, chairman of the new firm. Chief executive officer of the new company was Harry Vickers, an executive at Sperry Corporation. The intent was to merge two compatible firms. Sperry had a solid reputation as a successful supplier of technologies to the military, particularly during World War II, owned expertise in electronics, and conducted research in semiconductors. It had also dabbled with computers while doing research for the U.S. Navy. With the merger, Remington Rand changed. The two firms now grew momentarily and, on paper, in potential strength.

For further information, see: Franklin M. Fisher et al., *IBM and the U.S. Data Processing Industry: An Economic History* (New York: Praeger Publishers, 1983); Katharine D. Fishman, *The Computer Establishment* (New York: Harper & Row, 1981); Herman Lukoff, *From Dits to Bits: A Personal History of the Electronic Computer* (Portland, Ore.: Robotics Press, 1979); Stephen T. McClellan, *The Coming Computer*

Industry Shakeout: Winners, Losers, and Survivors (New York: John Wiley & Sons, 1984); Robert Sobel, *IBM: Colossus in Transition* (New York: Times Books, 1981); Nancy Stern, *From ENIAC to UNIVAC: An Appraisal of the Eckert-Mauchly Computers* (Bedford, Mass.: Digital Press, 1981).

S

SCIENTIFIC DATA SYSTEMS (SDS). SDS was a computer manufacturing company established in 1961 and acquired by Xerox Corporation† in 1969. The subsequent history of SDS is a major part of the copier giant's involvement in the data processing industry. The acquisition of SDS typified what happened to many small vendors in the industry. It was acquired for its successful products and marketing, bringing a special type of expertise to the acquiring organization. Conversely, companies seeking to expand their presence in the data processing community acquired specialized firms and then integrated their products and marketing into a larger whole. Acquisitions frequently answers the question of what happened to a company no longer in existence.

SDS was initially capitalized at $1 million. Its primary founder and president was Max Palevsky, a man who had first worked with computers in 1952 at Bendix Corporation. In 1956 he organized a computer manufacturing operation at Packard Bell. Seventeen people made up the entire employee roster at SDS when it introduced its first product, the SDS 910 computer (first shipped in mid–1962). The company enjoyed healthy growth rates in its early years, averaging compounded growth between 1962 and 1968 of 115 percent annually. In 1964 revenues were $20.5 million and in 1968 reached $100.7 million. The following year Xerox acquired the company with stock valued at $980 million. Xerox then changed SDS's name to XDS. Palevsky personally acquired $100 million in Xerox stock in this transaction.

The company's phenomenal success could be largely attributed to its early focus on real demand in two areas. First, there was a requirement for computers that could do real-time processing. Second, there was a growing demand for small and medium-sized computers in scientific applications. SDS sold products that satisfied each of these two markets and at a time when major vendors in the market were not delivering such products. As a consequence, profits remained very high. SDS's only competitor in the early 1960s was the Digital Equipment

Corporation (DEC)† which also sought to sell computers to engineers and scientific users.

The company's first important product was the SDS 910, a general-purpose computer that sold for $80,000 to $90,000. There were additional costs for peripheral equipment or process control devices (e.g., as was used on oil refining plants). The processor used paper tape and card equipment. The machine was built using existing components manufactured by other firms. Thus, for example, core memory came from Fabri-Tek and later also from Ampex. Tape drives arrived from Ampex and Computer Products, printers from the National Cash Register Company (NCR)† and Data Products Corporation.† Univac provided card punches, while Control Data Corporation (CDC)† built terminals. SDS designed some software and acquired other packages. As the decade passed, SDS manufactured peripheral equipment in-house while expanding its own development of software. All of these efforts were aimed at enhancing the 900 product line.

In 1963 SDS announced the 920, 930, and 9300, all of which were compatible with the 910. An SDS 930 was used at the Kennedy Space Center to simulate launch-related events as a training tool. The 930s competed against International Business Machines Corporation's (IBM's)† 1620 and S/360* and against Honeywell and DEC. The entire product line was sold primarily to scientific users. In 1964 it brought out the SDS 92, its first product to use monolithic integrated circuits. It was also one of the earliest computers in the data processing industry to rely on that technology. As was true of other computers based on monolithic circuits, manufacturing costs dropped when compared to those of earlier devices while reliability tripled. The SDS 92 was aimed at small users, and the SDS 925 at medium-sized installations. By the end of 1964 SDS had 1,357 employees. In order to raise money to fund its leased inventory and to conduct additional research and development, during 1964 SDS offered common stock for sale for the first time, generating $5 million in working capital.

Additional enhancements to its entire product line followed. For example, the 940, larger than the 930, came out as a time-sharing computer and sold for $1 million. In 1965 SDS introduced a software package called MANAGE for nonscientific users in need of a general decision-making application program. In 1965 SDS also nearly doubled the number of computers it installed.

The years 1964 and 1965 represent a watershed period in the history of data processing because it was during these years that IBM announced and shipped the S/360 line of computers. These machines ushered in a new generation of technology to which most computer vendors reacted, including SDS. It announced the Sigma Series in 1966 as its offering for third-generation computers and was aimed directly at IBM's products. The first machine was the Sigma 7 (announced in March 1966), followed by the Sigma 2 (announced in August). In early 1967 additional products emerged; by 1971 this family of computers consisted of the 2, 3, 5, 6, 7, 8, and 9. They were usable in both commercial and scientific environments. Like other vendors, SDS set prices for this equipment at between 10 to 15 percent below those for comparable S/360 processors. To provide adequate supplies of peripheral equipment to enhance the Sigma Series, SDS bought equipment from Potter Tapes, disk drives from CDC, printers from NCR, console typewriters from Teletype, while Uptime Company built card equipment. SDS also started to manufacture similar peripherals in-house. Software products were also introduced rapidly and in quantity, including compilers for the FORTRAN IV* and COBOL* programming languages.* Application software and utility packages complemented and broadened SDS's product lines.

SDS, therefore, rapidly became a fully integrated and important supplier of data processing equipment. Its new prominence was reflected in IBM's view of the company. In December 1964 T. V. Learson, a high-level executive at IBM and soon to be its chief executive officer, reported to the head of the company, Thomas J. Watson, Jr.,** that SDS was a major competitor for the IBM S/360 Model 40 and 50. At the time IBM considered it one of its top nine competitors. SDS continued to grow, doubling production in the critical period of 1966–1967. In 1969 its executives looked back with pride at the previous several years during which they had moved successfully from one generation of products to another, expanded manufacturing and research, while growing in revenues and profits. In 1969 SDS had $113 million in assets and 4,000 employees.

At that juncture, Xerox, interested in expanding its position in the world of data processing, negotiated for the acquisition of SDS. In 1968 Xerox had revenues of $896 million and net income of $116 million. Integrating SDS with Xerox made Xerox one of the top twelve manufacturers of computers. The subsequent history of SDS as XDS is that of Xerox.

For further information, see: Franklin M. Fisher et al., *IBM and the U.S. Data Processing Industry: An Economic History* (New York: Praeger Publishers, 1983).

SCS. See SOCIETY FOR COMPUTER SIMULATION

SDS. See SCIENTIFIC DATA SYSTEMS

SHARE. This organization is made up of users of International Business Machines Corporation (IBM)† computers. The name is not an acronym; rather, it stands for the fundamental principle for its existence: to share experiences in using such equipment. Over the years it has been an important user's group and has communicated to IBM the concerns of its customers regarding its computers. SHARE was also the first computer user group ever formed.

In the early 1950s IBM periodically brought together users of its products to discuss their usage and characteristics. Some users of such equipment, including the IBM CPC and later 701, also shared experiences among themselves. Then in 1954 several users of 701s in Los Angeles began to meet to discuss the best way to use such devices. The result of these meetings was a project called PACT, which sought to develop software that would cause the computer to do more of the work currently done by people. Then, in August 1955, IBM conducted a seminar in Los Angeles on the 704, and users next decided to meet on their own soon after, on August 22, 1955. This became the first meeting of SHARE. Twenty-two installations were represented and within a year had grown to sixty-two. By the end of 1980 there were over 1,500 installations in SHARE, and the number has grown since.

In the three decades since its formation, SHARE has launched a number of projects to improve software, educate its members, explain experiences with different devices and software, and advise IBM. It has established committees to discuss and examine key issues relevant to the data processing industry, and it has held annual conventions throughout the United States. In most years two meetings have been held and Proceedings published. Other publications have appeared, along with regional and topical seminars.

The organization's original founders were Lee Amaya (Lockheed Aircraft), Jack Strong, and Frank Wagner (both of North American Aviation). Its original membership was dominated by scientific users, although today it also includes business application members. Business users of data processing equipment formed their own users group, called GUIDE,† during the 1960s. SHARE focuses on such issues as software, database, interactive systems, operating systems, programming languages, and computer technology. It is a nonprofit organization, established as a corporation in January 1969.

For further information, see: Paul Armer, "SHARE—A Eulogy to Cooperative Effort," *Annals of the History of Computing* 2, no. 2 (April 1980): 122–129; T. B. Steel, "SHARE," in Anthony Ralston and Chester L. Meek, eds., *Encyclopedia of Computer Science* (New York: Petrocelli/Charter, 1976) pp. 1251–1252.

SID. See SOCIETY FOR INFORMATION DISPLAY

SOCIETY FOR COMPUTER SIMULATION (SCS). SCS, the oldest institution within the data processing industry, is dedicated to the encouragement of computer-based modeling and simulation. Its existence accents one of the most important uses of computers. Simulation, modeling, and "what if" analysis have frequently been the bedrock of much scientific research relying on computers. Almost all financial analysis, work in civil engineering, even projections of weather, flight paths for spaceships, and sales, represent variations of the same application.

SCS was formed on November 7, 1952, when thirty-nine individuals from thirteen organizations gathered together in Oxnard, California, to create a society dedicated to the use of computers in modeling. Many firms have been established in the region in order to manufacture aviation products and equipment for the armed services, all employing thousands of engineers and scientists. The driving force behind the creation of SCS was John McLeod, then with the U.S. Naval Air Missile Test Center at Point Magu, California. The organization was first named the Simulation Council. Another council, called the Midwestern Simulation Council, was established at the University of Michigan on June 28, 1954, with the same objective. The group in California then changed its name to the Western Simulation Council. The two merged and were incorporated on June 3, 1957, as a nonprofit institution called Simulation Councils, Incorporated (SCI). Then, in October 1978, it adopted its current name, believing it better reflected its mission.

By mid–1985 SCS had six regional councils within the United States and Canada, and one in Great Britain. The councils were geographical, reporting to the national headquarters in San Diego. Affiliations with other associations extended SCS's influence. The first such linkage was established on June 3, 1981, with the Dutch Benelux Simulation Society. Subsequently, during the 1980s other ties were created to the Association Française D'Intelligence Artificielle et des Systemes de Simulation (AFIAS) in France, the German Gesellschaft für Informatik—Department ASIM (usually called ASIM), and in Japan with the Japan Society for Simulation Technology (JSST).

These societies reflected a growing interest in the use of computers to do modeling. SCS encouraged such applications and sponsored a journal, *Simulation*, as a forum for discussing technical issues relevant to the society. *Simulation*, first published in January 1964, quickly became the major source of information about techniques and software for modeling. In 1971 SCS launched a collection of books, on average two per year, in its Simulation Series. In 1984 SCS began publication of the first in a series of *Transactions*, a collection of technical papers dedicated primarily to theoretical subjects. Other miscellaneous publications have appeared over the years. In addition, national conventions and seminars supported the primary mission of educating and encouraging members in their use of computers for modeling.

For further information, see: Charles A. Pratt, *The Society for Computer Simulation* (San Diego, Calif.: Society for Computer Simulation, n.d.).

SOCIETY FOR INFORMATION DISPLAY (SID). This association was founded in 1962, to quote a short brochure it published, to support those interested in the "presentation, exchange, and preservation of the ideas and technologies encompassed in" the use of information display. By 1985 its membership had approached 2,500 individuals. This organization sponsored conferences and published monographs and proceedings on relevant topics, much as did other such associations within the data processing industry. The centerpiece of SID's programs was the International Symposia, held annually with presentations and exhibits. By 1985 the representations by vendors reflected the products of some 100 manufacturers and service organizations. Tutorial lectures at these sessions were published as the *Seminar Lecture Notes*. The more detailed proceedings were published as annual volumes of the *Digest of Technical Papers*. Each year SID also sponsored the International Display Research Conference held in the

United States, Europe, or Japan. These sessions were technical and their results often appeared in SID's Proceedings.

SID's current structure evolved over the past twenty-five years from one to ten chapters. Eight chapters are in the United States and one each in Japan and Great Britain (with members resident in Ireland). Thus, like most other associations within the industry, SID's structure was national and somewhat international in scope. This association came into existence when members of the data processing industry realized they were a distinct industry with its own identity and needs. Americans have been characterized as tribal, forming more associations than people in other countries. SID, along with the Association for Computing Machinery (ACM),† Data Processing Management Association (DPMA),† and other associations within the industry came into existence almost at the dawn of data processing's major spurt of growth. They were motivated by the need to collect what little information then existed on new topics, to define requirements which were then passed on either to scientists or, as in the cases of SHARE† and GUIDE,† to vendors, while serving as a voice for a particular community, primarily within the American economy.

For further information, see: the monthly SID *Information Display Journal* and its newsletters, published out of the headquarters at 8055 W. Manchester Avenue, Suite 615, Playa del Rey, California, 90293.

SPERRY CORPORATION. See SPERRY RAND CORPORATION

SPERRY RAND CORPORATION. Sperry, as it was usually called, was a manufacturer of military electronics which merged with Remington Rand† in 1955 to become a major computer vendor. Its most famous contribution to the industry was a series of computers called the UNIVAC.* In the mid–1980s it continued to be an important supplier of large computers, but without the same influence it had had in the 1950s. Yet unlike many other entrants into the computer manufacturing business, Sperry frequently prospered and always survived, even though other electronics firms, such as Radio Corporation of America (RCA),† and General Electric (GE)† could not in the computer industry.

Sperry Corporation was incorporated on April 13, 1933, in Delaware as a holding company of other smaller firms, made up of 976 employees. Major elements of the company included Sperry Gyroscope Company, Ford Instrument Company, and Intercontinental Aviation, Incorporated. Sperry Gyroscope, incorporated on January 21, 1929, was the manufacturer of the gyrocompass, used by merchant marines and navies all over the world. During World War II it manufactured the gyroscope used in bombsights and torpedoes. The Ford Instrument Company, incorporated on November 15, 1916, made electronics gear for artillery control. These included rangekeepers, directorscopes, and mechanical computing devices primarily for antiaircraft fire control and artillery batteries. Intercontinental Aviation, incorporated on August 5, 1929, made aviation equipment outside the United States. During the 1930s sales hovered around $2.8 million annually. It acquired additional small companies producing aviation parts during the decade, positioning it well for sales to the military during World War II. Earnings had grown to $3.5 million in 1936, and by the end of 1939 gross income rose to $10.7 million on revenues of $24.4 million. Business boomed during World War II. In 1944, for example, sales were over $460 million, but, after the war, dropped sharply (in 1946) to $85.8 million because of decreased demand for its military products. During the Korean War, sales matched those of World War II, and, as in 1946, after the conflict ended they dipped again.

Although the majority of its sales had always been to the military, the company came out of World War II with some sectors of its business geared to commercial interests, including farm equipment. It had money to spend and, therefore, searched for an acquisition. The Remington Rand Company appeared to be a good candidate for complementing Sperry's operations. Sperry, on the one hand, had considerable technical and electronic expertise and excellent ties to the U.S. government. Remington Rand was a leader in the office marketplace and had a broad, loyal customer base. The two firms seemed a logical complement to each other, and thus they negotiated a merger. Each knew that Sperry could afford to invest heavily in research, and its leader, Harry Vickers, was willing to take on International Business Machines Corporation (IBM).† Remington Rand's customer set would make it easier for Sperry to sell products in the civilian market, and it could take advantage of Remington's sales force and marketing organization. Sperry's technology could be applied to civilian uses, including computers. Thus, in 1955 the two merged. Remington Rand became the Remington Rand Division of the Sperry Rand Corporation.

In addition to its aviation products, this new company could now claim to make typewriters, business machines, duplicator supplies, systems equipment (e.g., filing systems such as Kardex), adding and accounting devices, tabulating equipment, and electronic digital computers,* the most important of which was the UNIVAC I and, in 1957, the UNIVAC II.

It was thus ready to take on IBM which by then had seized the lead in computer-related developments and marketing. Yet the smart money was placed on Sperry Rand winning the battle for dominance in the office products arena. It seemed that UNIVAC's technology was in the lead; however, the sales force failed to perform effectively. The year of the merger saw UNIVACs dominating computer sales; yet within a year IBM had more computers installed or on order than Sperry Rand. Thus, the lack of an effective marketing force, or even additional speed in introducing new products, caused Sperry Rand to give computers less intensive attention than was evident at IBM. Sperry remained a formidable force, however. Using figures from 1963, for example, and in terms of dollars from data processing products in the United States, it was second to IBM, with revenues of $145.48 million. Its next competitor in size, holding third place, was American Telephone and Telegraph (AT&T)† which had sales of just over $97 million. IBM's were $1.244 billion.

In April 1964 IBM announced the S/360* family of computers, rocking the industry for the rest of the decade. For the first time, a vendor had introduced a major series of computers in which programs ran in various models with little or no conversion. Thus, customers could upgrade without major costly conversions of programs; such efforts might otherwise have cost more than the computers themselves. Such an announcement made customers confident that their investment in programming would be relatively secure. In addition to the hardware compatibility announcement on computers, peripherals were standardized to operate on multiple-size machines, and costs dropped dramatically. New and more reliable technologies were introduced. This product announcement created a technological revolution that ultimately led to growth for all companies. Sperry Rand, like other competitors, reacted to the announcement by deciding to introduce its own compatible line of computers called enhanced models of the 1040 and 1050, along with the 1107, 1108, and 1109 and a new round of price reductions to make all their products competitive with the S/360s. In 1965 the company's market share was still second to IBM's but nearly twice that of its next rival, Control Data Corporation (CDC)† and four times that of Honeywell or Burroughs.† Its gross revenues that year were $1.279 billion. In 1966 Sperry's computer operation finally had its first profitable year.

Throughout the 1960s the Univac Division, and thus Sperry Rand, was considered a mainframe manufacturer, that is, primarily a builder of large computers. It retained this status throughout the 1970s, a decade when mini- and microcomputers and distributed processing would usher in new companies that gained significant market share. Yet at the start of the decade Sperry Rand's prospects were bright. Radio Corporation of America (RCA)†—a major contender in the computer market in the 1960s—failed to make computer sales profitable. It was unwilling to commit vast new sums for research and development primarily to fight IBM and thus opted out of the market by selling its computer division to Sperry Rand in the fall of 1971. Sperry saw this acquisition as a means of penetrating the data processing industry even further. In 1973 it claimed that about 90 percent of RCA's customers remained with Sperry Rand and that in the previous year (1972), these same customers had acquired over $130 million worth of products. The company also worked to facilitate a technical merger of the two product sets (RCA's and Sperry's computers). By the end of 1974, 5 percent of RCA's old customers had installed Univac equipment and 77 percent still had RCA computers installed, generating rental revenue for Sperry Rand.

The years from 1964 onward were years of problems and successes for computer operations at Sperry Rand. Personnel turnover in the early 1960s became severe. Sperry Rand was selling several computer product lines instead of one, which enhanced costs and marketing complexity. In 1964, for example, it sold the UNIVAC II, the 1107, and the 490 systems. Older products had no successors ready to appear, and none of these three systems was compatible. Then IBM announced the S/360 family. Thus, when coupled with weak financial performance in the first half of the 1960s, Sperry Rand seemed to be facing potential problems.

Sperry Rand concentrated on sales to the U.S. government in the second half of the 1960s. As early as 1964, for instance, the U.S. Air Force ordered over 150 UNIVAC 1050-II systems along with other Sperry computers. Additional products were introduced which were important to the military, including the AN/UYK–5 and 7. The 490s were used for airline reservation systems. Important announcements included the 1108; yet software support failed to meet customer expectations. A third-generation computer was needed, and so Sperry Rand announced the 9000 series in two models to meet that requirement in early 1966. Subsequently, another model was announced in January 1968 called the 9400 which was first shipped in 1970. The Univac Division was a vertical operation, designing and manufacturing its own computers, peripherals, and software which it both leased and sold. R. E. MacDonald, president of the Univac Division (1966–1971) and also of Sperry Rand (1963–1971), pressed for new products priced about 10 percent lower than IBM's. The division did well under his stewardship, despite the company's lack of products as broad as IBM's. Data processing revenues grew in the U.S. market from $203 million in 1964 to $478 million in 1970. In 1969 the division represented the largest revenue generator within Sperry Rand. By the end of the decade, sales were being made throughout the world, with revenue growth overseas faster in many years than in the United States. The company thus entered the 1970s still in position to become a giant in a very crowded field of competitors.

In November 1970 the company announced the 1110, a general-purpose computer that could handle real-time, online systems, and batch applications. It could also support remote job entry and was intended for both scientific and business applications. One year later it brought out the 9700 and the 9000 series already mentioned. These products were introduced while RCA and Sperry Rand were negotiating the acquisition of RCA's computer division.

The acquisition was particularly exciting for the industry as a bellwether statement of Sperry's intent to remain a major force in the 1970s. By the terms of the agreement, Sperry made an initial payment of $70 million for RCA's computer operations and agreed to pay an additional $60 million which would come out of future revenues from the old RCA operations. Sperry bought over 1,000 customer installations in 500 accounts to which the Univac Division could sell its $130 million in products the following year. Sperry agreed to take over maintenance of RCA's equipment and to manage its leases. By the end of 1974 Sperry reported that it had gained $370 million in revenues from RCA's customers. In 1973 Sperry had acquired Information Storage Systems (ISS), which gave it greater capability to make disk storage equipment, adding strength and depth to the entire Sperry Rand product line.

Sales for the Univac Division remained strong during the 1970s. In 1970 revenues had totaled $635 million, but by 1980 they were at $2.319 billion. Although the largest portion of revenues in the early 1970s was generated by the 1100 series, other products soon contributed significantly to sales. In November 1976 the 1100/10 was introduced as a medium-sized general-purpose computer, and in June 1979 the 1100/60 became a competitor to the IBM 4341 range computer. The 90/60 and 90/70 computers were also important and were targeted as follow-on products to RCA's. Various models and enhancements appeared throughout the decade. Sperry continued to enhance its ability to manufacture disk drives by expanding ISS and acquiring other disk-manufacturing facilities.

Sperry entered the 1980s as a major vendor of mainframe computers. It was an era in which the percentage of all sales of data processing equipment coming from such devices was beginning to decline. Despite overall growth in sales, in 1980 Sperry was only the fifth largest supplier of data processing products, trailing behind such minicomputer manufacturers as Control Data Corporation (CDC) and Digital Equipment Corporation (DEC).† Then in 1982 and in 1983 profits dipped. At the same time the company dropped the word "Rand" from its title, calling itself Sperry Corporation—the name of one of the smaller portions of the firm back in the 1930s. It also stopped using the term *UNIVAC* after 1982, retiring one of the most famous words in the history of data processing.

Sperry's commitment to the mainframe market concerned officials within the company in the early 1980s. As in the past, there were moments in the company's history when technology did not advance rapidly enough (as in the 1950s), and it appeared that such a time was about to be repeated. In 1984 Sperry had about 6 percent of the mainframe market share versus 9 percent in 1975. Its 1984 share was still too high as a total contributor to Sperry's revenues, however, because the data processing industry was changing. It has been estimated that by the end of 1988 one-third of all data processing expenditures will go for either software or related services. Even though the industry experienced an annual growth rate of 10 percent in the 1980s, mainframe business did not grow proportionally.

There were yet other concerns. In retrospect, in the bountiful 1970s, the cost of maintaining two product lines—RCA's and Sperry's—was too great. It also seemed that the 1100/90 came out a year too late, and it was known that the company had suffered too much red ink with its minicomputer forays using the Varian minicomputer and the BC–7. Sperry's strategy in the mid–1980s, however, did stress its historical ties to military sales which were high in the United States under the Reagan administration. It was not uncommon for military sales to account for about 24 percent of the company's revenues. At the same time the company attempted to broaden its commercial electronics market share. It even introduced its own Personal Computer.

At Sperry's Computer Systems Division, marketing as opposed to finance or engineering dominated management for the first time by the mid–1980s. New managers from outside Sperry came in with new ideas, often cutting operating costs which had dug deeply into profits in the early 1980s. One immediate result was declining debt, which was previously a major source of problems for the company. In 1984 Sperry was about 30 percent capitalized, and management's target remained 25 percent. Earlier in the decade it had been higher. In the mid–1980s management conducted a serious reexamination of its product direction and strategy. Over the summer of 1986, it merged with Burroughs Corporation to form Unisys, the second largest vendor in the industry.

For further information, see: Franklin M. Fisher et al., *IBM and the U.S. Data Processing Industry: An Economic History* (New York: Praeger Publishers, 1983); Katharine D. Fishman, *The Computer Establishment* (New York: Harper & Row, 1981); Stephen T. McClellan, *The Coming Computer Industry Shakeout: Winners, Losers, and Survivors* (New York: John Wiley & Sons, 1984); Robert Sobel, *IBM: Colossus in Transition* (New York: Times Books, 1981); Nancy Stern, *From ENIAC to UNIVAC: An Appraisal of the Eckert-Mauchly Computers* (Bedford, Mass.: Digital Press, 1981).

T

TANDY CORPORATION. Tandy was best known for its electronic retail outlets under the name Radio Shack which, by 1986, had thousands of storefronts in North America and sold microcomputers and related peripheral equipment and supplies. Yet it was owned by a company that began as a supplier of components for shoes.

Dave Tandy and Norton Hinckley formed the Hinckley-Tandy Leather Company in Fort Worth, Texas, in 1927 to sell supplies to shoe repair shops in Texas. Through acquisitions, increases in sales and the extension of their product line after World War II to include leathercraft products, annual sales reached $1 million in 1948. The experience gained by the Tandy family in managing and operating retail outlets for their products influenced the company's future actions. Merchandizing and retailing provided a perspective that has governed the characteristics and policies of Tandy to the present, making it the largest retail supplier of data processing products for individual consumers by the early 1980s.

Retailing proved to be a successful path for the company. In the early 1950s the firm operated over 100 retail outlets exclusively for leathercraft products and, in 1952, annual sales reached $2.9 million. In 1955 sales amounted to $8 million, with earnings of $523,000, giving the firm financial strength. Following World War II the firm was known briefly as General American Industries which owned a variety of leather-related firms. A power struggle within the enterprise resulted in the Tandy family domination in 1960. Sales continued to grow and in 1968 reached $110 million. In 1961 the name of the company was changed to Tandy Corporation, and it operated 125 stores in the United States and in Canada. Thus, before taking any step into the field of electronics or data processing, the firm had a successful history of merchandizing and retailing.

Then in 1962, the chief executive officer of the company, Charles Tandy (son of the Tandy who founded the firm in 1927), decided that consumer electronics represented a crucial economic opportunity and bought Electronics Craft of Fort Worth, Texas. Encouraged by this acquisition, he sought a larger enterprise and settled on Radio Shack, a Boston-based mail order company that started in the 1920s as a supplier of electronic components for ham radio operators and electronics buffs. In 1962 its financial condition was poor, but it had nine retail outlets in the metropolitan area of Boston in addition to its mail order business, and sales reached $9 million. In April 1963 Tandy gained control of Radio Shack's management, and by June 30, 1965, his company owned 85 percent of its common stock, giving Tandy full power over Radio Shack.

To put the electronics firm on a sounder financial footing, Tandy forced it to sell off unprofitable inventories and to focus instead on fast turnover products with customer appeal. In the 1960s Radio Shack acquired the trademarks REALISTIC and ARCHER for its products. In the 1980s REALISTIC was still used on a wide range of small consumer products from portable tape recorders to television sets and stereo components. By the end of 1965 Radio Shack had sales of $20 million, with $4 million in profits. By the end of 1968 it was accounting for 40 percent of Tandy Corporation's total revenues.

In the 1960s the company also broadened its offerings from highly specialized electronic components to include consumer products: television sets, stereo components, and electronic gadgets for the home. By the end of 1968 it had 288 company-owned stores and an additional 28 leased facilities. Although Tandy still had its leather goods and even a nursery, the story of Tandy Corporation by the early 1970s was that of Radio Shack and electronics.

With sales of electronics still growing, Charles Tandy decided in 1976 that his company should begin moving into consumer data processing. The logical step was to explore the possibility of marketing microcomputers, which were just beginning to appear, particularly in electronics circles in California. Tandy felt that the time was ripe for adding this particular item to his product line. He felt strong because his firm had just virtually destroyed an arch competitor, Lafayette Electronics, which also owned retail outlets that appealed to the same market as Radio Shack. By 1976 Radio Shack so dominated the retail electronics market in the United States that for a time it was the target of a U.S. Department of Justice antitrust investigation. No legal action resulted.

Initially, Tandy Corporation did not want to manufacture microcomputers; it only wanted to resell someone else's in a few outlets. But the company had difficulty negotiating any arrangement. Then in December 1976, after initial probing of a potential design by Don French, a buyer for Radio Shack interested in micros, the company decided to develop its own machine. In August 1977 Tandy Corporation announced its new product called the TRS–80. It retailed for $399, and within a month the company had orders for 10,000 of them. In October Radio Shack opened its first store dedicated exclusively to computer-related products, establishing it in Fort Worth, hometown for Tandy Corporation. The store was a harbinger of things to come, not only for that company but also for thousands of stores run by Computerland, International Business Machines

Corporation (IBM),† and other vendors. TRS–80s were soon stocked in all of Radio Shack's stores, and other store outlets were opened by Radio Shack devoted solely to the sale of computer technology. Meanwhile, in November 1978 Charles Tandy, chief executive officer, died in his sleep. He had taken Tandy from a leather supplies business into the world of consumer electronics and retail computers. He personally made the decision to sell micros. Associates remembered his sarcastic humor, persuasive abilities, and unbounded energy, matched with his solid skills in merchandising.

The configuration of the TRS–80 which Tandy and his management team settled on was small compared to today's products. At the time, no one could foresee the public's reaction to the device, and forecasts of sales settled at 1,000 for the entire life of the product. This device, along with devices made by Apple,† Commodore, and Atari† created a worldwide demand for micros. By the mid–1980s over 2 million were sold worldwide per year by all vendors.

The original TRS–80 had limited capability because Radio Shack wanted to keep its price low, that is, within a range that would allow stores to move electronic products to consumers. Merchandising considerations more than technical issues governed the original configuration. The result was a processor whose configuration consisted of 4K of memory, a very limited BASIC,* all relying on a Z80 machine. It used cassettes to store data and had slow tape. The typewriter portion of the machine even lacked lower case letters. With sales rising so quickly, the company beefed up the machine with more function in its keyboard, a stronger BASIC richer in commands and power, additional memory, a printer, and disk storage. On May 30, 1979, Tandy announced the TRS–80 Model II which was a state-of-the-art device. A serious machine was introduced primarily because sales were high from the beginning, and the contribution to revenue and profits from these products was growing. Between 1978 and 1980, Radio Shack's microcomputer sales in North America rose from 1.8 to 12.7 percent.

New products appeared in 1980, including the Pocket Computer priced at $229. Its Color Computer cost $399 and introduced support for graphics and 16K of memory. The TRS–80 Model III also became available. Other machines on the market included new offerings from Commodore and Atari and even Apple—all pre-IBM Personal Computer (PC) technologies.

The first Radio Shack equipment did more to introduce the American public to micros than any other company's products. The early TRS–80s showed that a latent demand for such technologies existed that was profound in size and was an important encouragement to others to introduce similar products. And it may have helped those putting the business case together for IBM to enter that market—and, for that matter, may have encouraged nearly 150 other vendors to do the same.

Data processing products quickly boosted Tandy's performance. Net sales as a whole rose during the 1980s. In 1981 they were $1.691 billion and in 1983, $2.475 billion, leveling off only slightly in 1984 to $2.737 billion. Net income in that same period went from $169.6 million in 1981 to $281.9 million in 1984, giving Tandy a prominence in the data processing community. By then the company was known as Tandy/Radio Shack and had 5,928 company-owned stores in 1985 and an additional 3,090 franchised outlets in nearly 100 countries. No other electronics firm in the world had as many retail outlets. The company offered more than 800 computer-related products, mostly surrounding its microcomputers. It still sold stereo equipment, added telephonic gear, and had its more traditional electronic components in addition to a product line of various microcomputers.

In 1984 revenues from data processing amounted to $719.1 million, down 24 percent from 1983. Most industry observers felt that the problem was an aging product line in a market where over 100 vendors were introducing new microcomputers at lower prices with larger numbers of functions. Still, like other companies, Radio Shack introduced IBM PC-compatible products to increase the odds of selling equipment in a world now standardized on the large computer vendor's specifications for micros. In this new generation of post-IBM micros, Tandy introduced the Tandy 1000 in December 1984. Most competitors, however, had brought out their new round of products as early as 1983, putting Tandy behind many other firms going into 1985.

Despite these difficulties—and they were also evident elsewhere within the data processing community—Tandy's experience mirrored patterns of behavior common in the industry since the 1950s. First, it began as a company with no interest in data processing and, second, it acquired the interest through purchases and good strategic decisions. Its management saw the potential of electronics—not data processing—but later the value of data processing products. Fourth, it grew not only by expanding existing services, in this case storefronts, but also by relying on acquisitions. Finally, many vendors, and not just those selling desktop computers, had learned a lesson: one had to introduce products with either perfect timing or at an earlier time than others, keeping them current and cost-effective. To slip by even a half year threatened severe and rapid financial punishment. The turbulent world of microcomputer sales (either for home or office markets) was always volatile and crowded with competition. Tandy's experience reflected these circumstances.

For further information, see: Paul Freiberger and Michael Swaine, *Fire in the Valley: The Making of the Personal Computer* (Berkeley, Calif.: Osborne/McGraw-Hill, 1984); James L. West, *Tandy Corporation: "Start on a Shoe String"* (New York: Newcomen Society in North America, 1968).

TELEX. Telex earned its place in history as much by filing a lawsuit as by selling products that were plug-compatible to International Business Machines Corporation's (IBM's)†. It began in the 1930s as a firm selling hearing aids and acoustic products, and was a family-owned business. It was not until 1959, then under new management, that it broadened its offerings in the general field of electronics. In the early 1960s these included parts, instruments, and control equipment. It even manufactured phonograph players and radios for sale in the United States.

Telex's initial flirtation with data processing came in 1962 when it acquired Midwestern Instruments, a company that built electromechanical equipment and telemetric devices, both of which were used in the American space program. Telex also bought the company because it made magnetic tape and sold tape drives to companies within the data processing industry. Telex's major role as a vendor within the industry came when it decided to sell tape drives that were plug-compatible to IBM's, beginning in the second half of the 1960s. This decision was made by Stephen J. Jatras, the company's president, who saw a huge market already made up of 53,000 IBM tape drives installed by the start of 1966.

Jatras's company elected to make equipment that functioned like IBM's but sold for less. Its first contract came from DuPont to replace some IBM 729 tape drives, giving the company the incentive to finally announce its own products. Telex's version of these first appeared in 1967. Business proved better than anticipated, and so the company next offered disk drives and printers, also aimed at IBM. Information Storage Systems (ISS) and Telex signed an agreement on April 21, 1969, whereby Telex would sell and maintain IBM 2311-type and IBM 2314-type disk drives made by ISS. The following year, Telex expanded its compatible strategy by coming to an agreement with Control Data Corporation (CDC)† to sell CDC's printers. These activities brought Telex's revenues from data processing sales within the United States up from $870,000 in 1966 to over $23 million in 1969; it closed 1970 with sales of $65,628,000. In 1971 revenues reached $81 million, generating profits of $5.5 million.

In the early 1970s numerous lawsuits were filed against IBM by various plug-compatible (PCM) vendors who accused the company of monopolistic behavior. Throughout the decade, IBM fought them and won. CDC filed one of the first, in 1968, but settled out of court in 1973. CDC's business partner, Telex, entered the fray in 1972, and after an initial trial finding in favor of Telex, the case was overthrown by a court of appeals with a judgment in favor to IBM. Telex attempted to appeal to the U.S. Supreme Court but settled with IBM before that option could be exercised.

The case gained an enormous amount of publicity in the early 1970s, becoming one of the major events of the decade within the industry. Telex had asked for damages amounting to $416 million which, when trebled as per normal practice in such cases, would total $1.2 billion. In January 1973 IBM countersued for infringement of patents and copyrights, also alleging that trade secrets had been stolen. The initial court found both companies guilty of the charges made by the other on September 14, 1973. IBM was accused of having gone after the plug-compatible industry in an attempt to destroy it. In turn, Telex was fined for abusing IBM's patents and copyrights. Telex's initial victory led nearly a dozen other companies to sue IBM in hopes of also winning treble fines—an amount far in excess of what they would have made through normal sales of products. For several years, therefore, suits against IBM appeared to be a more profitable way to run a business, but all failed to win even once, incurring instead heavy legal expenses.

The reversal in the Telex case came in 1975 when all fines against IBM were dismissed and the large company was vindicated of any wrongdoing. Telex's fines were maintained. In October, unable to continue the lawsuit owing to financial exhaustion, Telex negotiated a settlement with IBM whereby Telex would not pursue the case any further and each agreed not to pay the other any money. This turn of events did discourage some of the other litigants from carrying their cases any further.

Meanwhile, business continued. Telex reacted to IBM's announcement of the S/370* family of computers in 1970–1971 by bringing out compatible tape drives to sell against the IBM 3420. IBM announced its tape drives in November 1970, and Telex in December, calling its product the 6420 tape drive and the 6803 tape controller. It had even hired away from IBM the engineer who had developed IBM's 3803 controller, Howard Gruver, to develop its plug-compatible product. Shipments of the tape drives began in November 1971 and proved acceptable during 1972 with some 2,000 units shipped. Telex also launched a rate war against IBM. Yet Telex experienced difficulty in marketing CDC's disk drives which were aimed at IBM's 3330s in the early 1970s, blunting some of the sharpness of its attacks.

From 1970 through 1973, Telex suffered a variety of manufacturing and development problems with its 6420 product line; problems also surfaced with the disk drives. Quality and manufacturing concerns continued to plague the company when a new firm came into the plug-compatible market called Storage Technology Corporation (STC). It aggressively and quickly displaced Telex's products in 1972 and 1973.

Telex's sales in 1974, although higher than in 1971, reflected another year of losses rather than profits. In 1975 the firm logged in a profitable year with sales reaching $106.1 million. From then on, its financial health improved slowly, and in 1979 it closed out the year on revenues of $148.2 million. Its better condition came as a result of more products, many of which came through acquisition of other assets. Thus, in December 1976 it acquired the division that made terminals at United Technologies Corporation. By the end of the decade, this division had made a product for Telex called remote access communications terminals (REACT). In January 1977 Telex acquired Gulliver Technology which led to the introduction in 1978 of a tape drive that could operate at 6,250 bits-per-inch (bpi), representing the densest packing of data on a tape drive in the industry. However, IBM had introduced this technology before Telex; Telex simply mimicked it as part of its plug-compatible strategy. Its device was aimed at the IBM 3420 Models 4, 6, and 8. In 1978 Telex bought General Computer Systems (GCS); this acquisition gave it the capability of using the GCS 2100, a system for supporting terminals linked to a processor that used magnetic disk and tape drives, card peripherals, telecommunications, printers, and other components. The 2100 was Telex's entry into the rapidly growing distributed processing market.

Telex came out of the 1970s in relatively weakened shape as a result of its battles with IBM both in the courts and in the market. STC and others had acquired a position in the PCM arena at Telex's expense. Yet Telex held on in the early 1980s. Management changed, and for a short period, it withdrew from the PCM market. Debts were also a nagging problem. Telex did not lose money in 1980, nor did it make a profit. It elected to build its terminal business and stay away from disk and tape drives in an attempt to recover. By the end of 1983 it had sold $320 million in goods which generated profits of $34 million. Its products were compatible with IBM's widely used 3270 family of cathode ray terminals (CRTs). One industry observer estimated that in 1984 Telex had 9 percent of the market for CRTs, IBM closer to 55 percent, and Raytheon† nearer 15 percent, while ITT's (Courier) products had 12 percent. The demand for CRTs rose during the mid–1980s by nearly 45 percent annually, although with declining profits per unit. In the same period, Telex began to diversify its product line again as a plug-compatible vendor.

Telex turned in a performance that was widely varying but not any wilder than that of many other firms in the industry. Its growth rates in the 1960s were impressive even by industry standards. In the early 1970s, Telex experienced modest successes but almost went out of business in the second half of the decade. Success in the early 1980s was remarkable inasmuch as high-technology firms that undergo several difficult years are seldom able to recover. Although it remained a small vendor within the industry (as of 1986), it was one of the oldest suppliers of peripherals in the business.

For further information, see: Franklin M. Fisher et al., *Folded, Spindled, and Mutilated: Economic Analysis and U.S. v. IBM* (Cambridge, Mass.: MIT Press, 1983) and *IBM and the U.S. Data Processing Industry: An Economic History* (New York: Praeger Publishers, 1983); Stephen T. McClellan, *The Coming Computer Industry Shakeout* (New York: John Wiley & Sons, 1984); Robert Sobel, *IBM: Colossus in Transition* (New York: Times Books, 1981).

TEXAS INSTRUMENTS, INC. (TI). TI is one of the largest U.S. electronics firms, headquartered in Dallas, Texas (helping to establish a nickname for the area: Silicon Prairie). It is a major producer of integrated circuits (otherwise known as chips*) for the data processing industry. TI marketed some of the earliest and most widely used hand-held calculators, dabbled in computer products, and became a large supplier of components for computer-related hardware across the entire industry beginning in the 1950s. Yet its beginnings were in oil-related services.

The original company was founded in 1930 when J. Clarence Karcher and Eugene McDermott formed a partnership called the Geophysical Service (later renamed Geophysical Service, Inc. [GSI]). These two young scientists wanted to explore for oil using technologies advanced for their day. In their case they used methods that measured seismic waves to map conditions under the surface of the earth to determine whether the indications suggested deposits of oil. They developed devices to help search for oil and were successful throughout the 1930s. At the start of World War II the company explored and drilled. The founders chose to split its two functions, naming the oil division Coronado Corporation and placing GSI under it as a subsidiary. In 1941 the oil company was sold off, and on December 6, 1941, GSI was bought by several of its managers. During World War II the company used its expertise in electronics to carry out projects for the Signal Corps and to build airborne magnetic detectors. One byproduct of this experience was close relations with military agencies buying electronic hardware.

Following the war, a veteran from the U.S. Navy, Lieutenant Patrick Eugene Haggerty, joined GSI and convinced management to expand its production of electronics. It did so, providing the U.S. Navy with airborne radar and sonar systems. Sales of electronic equipment to the U.S. government generated rapid growth in revenues. By the end of 1947 the company employed 643 people and had revenues of $3 million. By the end of 1949 it had a population of 792 and sales of $5.7 million.

The company changed its name in 1950 to General Instruments, Inc., with GSI a subsidiary, a change reflecting the growing importance of electronics for what was otherwise an oil exploration firm. But because the company's name was too similar to another's, the following year management changed it to Texas Instruments, Inc. In early 1952, America Telephone and Telegraph (AT&T),† which had developed the transistor at its Bell Telephone Laboratories,† signed an important agreement with twenty companies, including TI, to manufacture these components. That year sales for all products and services reached $20 million, and TI received its first order for transistors—ten in fact—in December. No product was as closely associated with TI during the 1950s as the transistor, which it sold aggressively to many manufacturers of data processing equipment during the following decade as well as to other vendors of electronic products.

TI also grew through acquisitions made in 1953. One of these, the Houston Technical Laboratories, specialized in gravity meters used in geophysical work. Its acquisition represented TI's renewed commitment to geodetic explorations. During the 1950s TI worked in Saudi Arabia and in Africa as well as in the United States. It also launched a seismic ship called the *M/V Sonic*, the first of many such ships built and manned by TI over the next two decades. In 1953 TI established its Central Research Laboratories, which developed many data processing-related items in future years.

During the mid–1950s the company decided to manufacture transistor-based products for the general public, beginning with a radio called the Regency. In 1954 the company announced that it had the first commercially produced silicon transistors, reflecting a growing interest in TI with semiconductors and what eventually would be called the chip.* TI's interest was shared by other companies. The U.S. military was very interested in miniaturizing electronics and, by the late 1950s, especially for rockets and satellites. Since the government was perhaps the largest buyer of electronics in the United States and the military the most important buyer, its wishes always translated into products delivered by manufacturing companies. Its encouragement of new electronics through acquisitions or through sponsorship of research projects leading to miniaturization directly influenced companies. In addition, electronics firms that were relatively young saw the trend toward smaller packaging of electronics as a way of providing less expensive, more reliable, and thus competitive products to put up against the likes of such giants as Radio Corporation of America (RCA)† and General Electric.†

While major U.S. electronics firms were being sued for price fixing by the American government in the early 1960s, young companies were competing for new markets. TI's story was part of that canvas of changing patterns within electronics. In the 1950s Fairchild Semiconductor Corporation† was developing the chip concurrently with TI. At the Texas-based firm, Jack Kilby** built one in 1958 or early 1959, while Robert Noyce** did the same at Fairchild. In July 1959 Noyce applied for a patent at almost the same time as TI. As a result, TI sued Fairchild for infringement of its patent to the semiconductor process. The case lasted several years, with both the electronics industry and the data processing community concluding that the two engineers had invented the integrated circuit. Kilby was inducted into the Inventors' Hall of Fame as the prime inventor, but historians later concluded that Noyce also contributed by dramatically improving the quality of integrated circuits to the point of making them commercially viable.

Seeing the advantages of integrated circuits, TI did not hesitate to manufacture chips for industrial purposes. The company strengthened its hand the same year it filed for patent by acquiring Metals and Controls Corporation of Attleboro, Massachusetts. This company built electrical controls and had skills in metallurgical fabrication. TI entered the 1960s with 17,000 employees and annual sales of $233 million. It built more electronics for the military, including terrain-following radar used by tactical aircraft and a night vision airborne infrared system, contributed to U.S. air-to-ground missiles, and provided the integrated circuits used on the Minuteman ICBM. Although it continued to develop new technologies for its geodetic piece of the business, by the early 1960s electronics contributed 80 percent of all sales.

Oil exploration continued to occupy TI as much as it had since the 1940s. The company applied its growing expertise in electronics to this field, including the application of data processing in general. As early as 1953, a design engineer named J. Fred Bucy worked on a system to record seismic data in digital rather than analog form called the Digital Field System (DFS), using a processor named the Texas Instruments Automatic Computer (TIAC). It was one of the first computers built almost entirely out of transistors. It was not TI's first adventure into data processing, for TIAC replaced *seisMAC*, a system that processed analog data using a digital program in a specialized processor. Throughout the 1960s and into the 1970s, TI built specialized data processing equipment to handle seismic applications. One of the more important systems was the Advanced Scientific Computer (ASC), constructed in the 1970s, which was an early supercomputer. TI used it for its own purposes, and the U.S. National Oceanographic and Atmospheric Administration (NOAA) used it to do research on weather.

Meanwhile, the sale of electronic components, and chips in particular, supported sales throughout the 1960s. By 1969 TI had revenues of $800 million, and although it faced stiff competition in the manufacture and sale of chips, it moved into a different field of data processing with the introduction of the hand calculator. In 1971 Michael J. Cochran and Gary Boone, both at TI, developed the "calculator-on-chip." The first product for consumers using this technology was the DataMath, a hand-held calculator introduced in 1972. When it was test marketed in Dallas that year, it sold at Neiman-Marcus and at Sanger-Harris department stores for $249.95. By the early 1980s the same function was available on TI calculators for less than $12. This particular calculator was the first such device most Americans ever saw or owned. Others, such as those built by Hewlett-Packard,† sold for hundreds of dollars more and in fewer numbers to scientists and engineers. TI's was meant for use by anyone, regardless of expertise. It became popular because it could add, subtract, multiply, and divide and, like today's devices, display the results. The difference is that the gadget was almost an inch thick and filled a hand, while today's instruments have shrunk to the size of a plastic credit card and have considerably more function.

Hand-held calculators sold well, and in 1974 TI introduced the TMS1000 microcomputer which could be used to make other calculators and data terminals. This component provided nearly half of the world's supply of single-chip microcomputers by the late 1970s. In 1974, TI also introduced other hand-held calculators, including the SR–50 and SR–51, for scientific applications, while the TI–5050 became the first portable, battery-operated printing calculator. In 1975 the company brought out a digital watch followed soon after by many other firms which almost as an entire group (including TI) lost millions of dollars. Their problem was due to prices falling rapidly for these watches, prices that dropped quicker than manufacturing costs. The first TI digital watch, although expensive, dropped to under $20 by 1976; the same function sold for less than $10 in 1977.

Competition for the sale of chips became keen in the early 1970s within the United States and, by the end of the decade also came from Japan. For example, in December 1971 Intel† announced its 4004 chip, and, in March 1972 TI brought out an 8-bit chip. Within a month Intel had its own 8-bit microprocessor, called the 8008. Unfortunately for TI, its 9-bit processor was more expensive than Intel's and three times as big. TI sought to use its chip in calculators, while Intel expanded its share of the semiconductor market by selling the 8008 for use in a broader set of applications.

Nonetheless, TI continued introducing new chips and products throughout the

1970s. In 1976 it had bubble memory which, as of 1986, had yet to become a popular technology. The TI–58 and TI–59 hand-held programmable calculators came out in 1977 and made money for the company. In the late 1970s TI manufactured voice synthesizers and children's calculator-like toys based on chip technologies. In 1979 annual sales for all products and services reached $3.2 billion. The company also experimented with commercial computers, including its DS 990 Commercial Computer System and the TI Series 700 Distributed Processing Systems. As with many other intermediate-sized computers, they had a normal complement of features. For instance, the 990 could be configured with up to 2 million bytes of memory, had a database management system,* and supported three widely used programming languages: COBOL,* FORTRAN,* and BASIC.* TI also built or sold peripheral equipment, including disk and tape drives, card I/O, and printers, to support these systems.

TI elected to enter the personal computer market early with disastrous results. During the first nine months of 1983 it lost $550 million based on sales of $400 million and temporarily left the market before the end of the year. As with digital watches, TI's management had anticipated that a low-priced personal computer would buy market share, which in turn would permit economies of scale in manufacturing to make these products profitable. Although the strategy had usually worked in the manufacture and sale of semiconductor chips, it failed with microcomputers. Simply put, retail prices for personal computers dropped faster than manufacturing expenses; TI faced this same problem with hand-held calculators, a market it once dominated. As the number of competitors in a market increased, so too prices dropped; TI was simply proof of that axiom at work in the data processing industry. In each case the results for TI were generally unfortunate. Revenues gave some indication of the tough situation. In 1984 the company booked $5.741 billion in sales, of which only $860 million came from data processing products. In 1983, the year of the personal computer problem, data processing revenues had been only $800 million.

The year 1984 was a year in which the entire semiconductor industry experienced a slump in sales, although the government accused TI of not always producing high-quality microcircuits. For TI overall revenues were up some 25 percent over 1983. The company had begun to develop software for artificial intelligence,* some of which ran on either its personal computers or those of other vendors. During 1984 sales of software generated $40 million, up 33 percent over the previous year, which suggested that a new chapter in the history of TI was possible.

For further information, see: T. R. Reid, *The Chip: How Two Americans Invented the Microchip and Launched a Revolution* (New York: Simon & Schuster, 1984); E. M. Rogers and J. K. Larsen, *Silicon Valley Fever: Growth of High-Technology Culture* (New York: Basic Books, 1984); Texas Instruments, Inc., *50 Years of Innovation: The History of Texas Instruments, A Story of People and Their Ideas* (Dallas: Texas Instruments, 1980).

TI. See TEXAS INSTRUMENTS, INC.

TRW. Within the data processing community TRW is best known for building space-related products (such as satellites and spaceships), for operating one of the largest credit-checking systems in the United States, and for building components used in data processing products made by other vendors. TRW has also been a leading manufacturer of automotive and aircraft components during most of the twentieth century.

The company was founded in 1901 as the Cleveland Cap Screw Company making cap screws, bolts, and studs. In 1904 it began manufacturing automotive valves, and in 1908 it was renamed the Electric Welding Company. In 1915 the firm's name changed to the Steel Products Company, and it became the largest manufacturer of automotive valves in the United States. During World War I it produced valves for automotive equipment as well as for French and American military aircraft. During the 1920s the Silcrome valve and other automotive products dominated the company's offerings. In 1926 it changed names again, this time to Thompson Products, Inc., after its president, Charles Thompson (president, 1915–1933). By the time he died in 1933, the company was the world's largest producer of automotive valves. During the 1930s sales slumped along with profits, yet the company expanded its sales of aircraft valves and other parts. This success was due in large part to the view of its president, Frederick C. Crawford (president, 1933–1953, chairman, 1953–1958), that aerospace represented another growing market. During World War II its energies were directed to the manufacture of war-related products for the U.S. government, particularly engine parts for aircraft and mostly at its Tapco facility in Cleveland, Ohio. After the war sales continued to grow. In 1946 they were $64 million and in 1950, $125 million. During the Korean War the company repeated its experience of World War II along with growth in sales; it closed out 1953 with revenues of $327 million.

In 1953 the company sponsored the creation of a small firm in Los Angeles by two engineers named Simon Ramo** and Dean Wooldridge,** called the Ramo-Wooldridge Corporation. Its purpose was to provide systems engineering or integration of multiple components for large projects initiated by the U.S. Air Force. Its most important early contract came in 1955, calling for it to serve as technical adviser to the U.S. Air Force in the construction of ballistic missiles, a program funded with $17 billion. Within two years the company had responsibility for coordinating the efforts of 220 prime contractors and several thousand subcontractors, generating $28 million for the firm. In 1958 it merged its 3,000 employees with the old Thompson Company under the name of Thompson Ramo Wooldridge. In 1965 the company adopted the name it uses today—TRW—even though it had been used informally before. TRW spent the rest of the 1950s and all of the 1960s associated with various projects sponsored by the U.S. Air Force, including subsequent improvements in ballistic missiles.

During the 1950s the company also acquired other products and facilities, which included Ross Steering gears, Marlin-Rockwell bearings, and United-Carr fasteners, all to bolster its automotive offerings. TRW also expanded its operations in aircraft engineering, becoming the first company to launch a spacecraft. Done in 1958, it was called the Pioneer I and became the first of some 175 satellites built by TRW between 1958 and 1984. By the early 1980s TRW was producing about one-third of all unmanned spacecraft.

Like many other young firms in the business of data processing in the 1950s and 1960s, TRW depended heavily on government contracts for survival. During the 1970s about 70 percent of all its sales came from the U.S. government despite its strong presence in the automotive industry. Meanwhile, it was building sales operations around the world as part of its attempt to diversify while its engineers became increasingly involved with data processing's technology as part of the company's growing role in U.S. space programs. By 1977, when TRW turned in revenues of $3.2 billion, government contracts accounted for only 20 percent of its business. In large part this was due to the decision made by its president, Horace A. Shepard (president, 1962–1969; chairman, 1969–1977) to reduce ties to the government and to expand into other fields, including more formally into data processing.

In the 1970s TRW introduced itself to the manufacture of chips* and in the 1980s implanted such technology in automotive parts. It also used this technology in satellites and other telecommunications gear during the 1970s and 1980s. TRW built Pioneer 10 which traveled past the Sun for some 2.3 billion miles. This same device passed by all the planets into deep space during June 1983, making it the first manufactured vehicle to travel outside our solar system. This device received so much public attention because it had a brass plate mounted on it depicting the shape of a human and had coded on it sounds of earth, human speech, music and mathematical symbols. TRW also served as a prime vendor for the Apollo lunar module.

TRW established one of the largest computerized credit-checking services in the United States that contained files on over 90 million people by the end of 1984. Despite a brief scandal in the early 1980s in which these files were penetrated by unauthorized users, thus fueling a growing concern within the data processing community about how best to manage data security, it remained a growing and profitable venture which in 1984 contributed $100 million in revenues.

The company also wrote software on a contract basis and during 1984 produced 10 million lines of code. It claimed that only International Business Machines Corporation (IBM)† wrote more. In 1983 this part of TRW's business contributed $760 million in revenues and in 1984, $825 million out of the total company sales of slightly over $6 billion. By the end of 1984 all data processing revenues contributed 18 percent of TRW's sales and grew proportionately in 1985 and 1986. This trend in turn made TRW a major firm within the data processing community. *Datamation*, a widely read journal within that industry, ranked it twenty-fifth from the top in 1984 in terms of dollar sales. That ranking put it ahead in size of such better known firms within data processing such as Northern Telecom, Ltd. (NTL)† (twenty-seventh), Automatic Data Processing (ADP)† (twenty-ninth), Texas Instruments† (thirty-second), Amdahl† (thirty-sixth), Prime Computer† (fortieth), and even National Semiconductor (forty-seventh).

For further information, see: TRW, *The Little Brown Hen That Could* (Cleveland, Ohio: TRW, undated [1983?].

U

UNIVAC. See SPERRY RAND CORPORATION

USE. This organization is made up of data centers that have had UNIVAC* computers installed. Like nearly a dozen similar associations, USE represents a band of people using a particular vendor's equipment, seeking to share experiences and information concerning commonly used computers and, furthermore, to relay requirements back to the manufacturer of such systems. The name grew out of UNIVAC Scientific Exchange (USE). It came into existence in December 1955 when four customers about to take delivery on UNIVAC 1103A computers met with representatives from UNIVAC in Los Angeles. In the late 1950s committees were established to write standards, advise on programming, and produce publications on UNIVAC's equipment. It wrote the USE language and a compiler for it, and played a role in the definition of requirements for subsequent computers. In 1966, reflecting the growing size of the organization, USE dropped the word "Scientific" from its title.

USE is one of two associations serving the needs of UNIVAC's customers. The other is the America's UNIVAC Users' Association, first known as the UUA and later as the AUUA, Inc. In 1961 the UUA and USE began negotiating as to how best to cooperate and, beginning in 1965 and through 1968, held joint conferences. After 1968 they no longer sponsored joint annual meetings. In 1980 USE, Inc., had 375 organizations as members and already had established a pattern of holding two annual conferences typically attended by about 1,000 people. These are important events, serving as the main avenue for carrying out USE's mission.

For further information, see: M. M. Maynard, "USE," in Anthony Ralston and Edwin D. Reilly, Jr., eds., *Encyclopedia of Computer Science and Engineering* (New York: Van Nostrand Reinhold, 1983): 1552–1553.

V

VERBATIM CORPORATION. Although little publicized, Verbatim became the largest manufacturer of floppy disks in the data processing industry. Floppy disks became the primary medium for the transportation of data and programs in microcomputers and in many minicomputers. In addition, since Verbatim was founded in 1969, it also had the distinction of being a relatively old company within the world of computers.

Verbatim was founded by Reid Anderson, (1917-). At the age of forty-six Anderson established his own company for the purpose of making and selling transistorized metronome/tuners for musicians. He had gained considerable experience in research and development as an employee of Bell Labs† for seventeen years during the 1940s and 1950s which he used in his commercial ventures. In 1956 he became director of physical research at the National Cash Register Company (NCR).† Within eighteen months he left, moving to the growing heart of a new generation of computer-related companies—Palo Alto, California—taking a job with the Stanford Research Institute (SRI).

After several years of making and selling the Tempo Tuner in the 1960s, Anderson began manufacturing display terminals and developed an acoustic modem for moving digital data across telephone lines. Splitting up with a partner gave him the opportunity to establish Information Terminals Corporation in 1969, the company that would later be renamed Verbatim. With $300,000 in capital, he began to develop magnetic tape-based computer products. These were all cassettes which, as a medium for storing machine-readable data, proved unattractive. Then, in 1973 International Business Machines Corporation (IBM)† introduced the floppy disk, which looked like an odd-sized 45 rpm record. Verbatim signed a contract with IBM to manufacture floppies (as they were called), and it was that product which became the mainstay of the company.

Throughout the 1970s and 1980s, Verbatim acquired manufacturing rights to other floppy-related products, usually other floppy disks of different sizes, which it manufactured for various companies. The firm dominated the new market for floppy disks. In 1979, when it went public, the offering was totally sold out. By 1984 the firm had 2,800 employees, annual sales of some $120 million, and profits at $14 million. Although the company experienced severe quality problems with its products in the late 1970s, in the 1980s these were behind it. Verbatim introduced the Datalife, the first floppy disk in the industry to have a five-year warranty. In summary, the company relied largely on its expertise in a particular area to provide a product while relying on the technological developments of others to thrive.

For further information, see: Robert Levering et al., *The Computer Entrepreneurs* (New York: New American Library, 1984).

VIATRON COMPUTER SYSTEMS CORPORATION. One of the first companies to produce a microcomputer, Viatron survived only one year before going into receivership. Its history set an early pattern for many companies that flourished for only a brief time either because their products were too advanced for the market or because of bad management by bright inventors.

In March 1968 a scientist at the Mitre Corporation, a leading center for research on computerized technology in California during the early to mid–1960s, left the firm with thirteen colleagues to form a new company. Dr. Edward M. Bennett, head of Viatron, announced their first product in October 1968. It was a small computer that could communicate with a host processor remotely and rented for $39 a month. His idea was to keep the price low in order to encourage mass production, which in turn would generate sufficient sales to justify the price. This product, intended for the general public, sought to do what International Business Machines Corporation's (IBM's)† PC and Apple's† machines did over a decade later: offer cheap computing for the masses. By the start of 1969 the company had acquired 14,000 orders for its product and began to operate in the red. Delivery, originally scheduled for that summer, came in October with Viatron's outlay already at $9.4 million. The company next announced two larger models, one of which would rent for $99 per month.

The company expanded into new facilities, hired people, and launched an aggressive advertising campaign. By April 1, 1970, the company's backlog of orders was valued at $100 million, with one order amounting to 34,000 terminals/computers. It had $36 million in credit lines, could manufacture 600 units per month and had ordered $63 million worth of chips.* The company was already the subject of considerable and favorable press coverage within the data processing industry and was formulating plans for manufacturing additional units in the Far East.

Within thirty days, however, problems emerged. Costs were found to be too high, lead times for machines ran so long that customers began to cancel orders, and the U.S. economy began to slip into a recession. Viatron reacted by doubling its prices. Layoffs began, while the total number of terminals shipped year-to-date had been around 1,500. *Computerworld*, the industry's leading newspaper, began to question the viability of the company. The demise of the company came quickly. In July Bennett was fired by his board of directors. Layoffs continued while the board looked for a buyer. By the end of the year, the company had seen three presidents, had no backlog, and owed $15 million in interest charges on debts. Viatron officially went bankrupt in the spring of 1971 and into receivership with only seventeen employees. At its peak, in the Spring of 1970, it had nearly 1,000 employees. In 1974 the company ceased to exist.

Viatron's demise was the outcome of several events. First, Viatron's dealers refused to carry the product line when the company withdrew leases in its attempt to reduce cash outflows. The cost of parts, especially chips, exceeded anticipated budgets and thus was a second reason for its ill health. Third, the company brought out a product before its time, a cardinal sin that ruined other companies within the data processing industry. The market was not prepared to accept the mass quantities of such devices that it would be in the late 1970s and early 1980s when companies such as IBM and Apple would build, ship, and have demand for hundreds of thousands of such products within an average year. Also fatal was the company's inept management which was unable to control and manage growth in a logical and organized manner.

For further information, see: Franklin M. Fisher et al., *IBM and the U.S. Data Processing Industry: An Economic History* (New York: Praeger Publishers, 1983); Jack B. Rochester and John Gantz, *The Naked Computer* (New York: William Morrow & Co., 1983).

VIM. This organization, made up of users of computers manufactured by Control Data Corporation (CDC),† began as the "VIM Organization for Control Data Corporation (CDC) 6000 Series Computers." Its first meeting was held in 1965 with attendees from ten organizations. The name VIM represents a pseudo-Roman numeral 6,000. In March 1970 this association organized as VIM, Inc. In October 1979 membership expanded to include any user of CDC's computer products, not just the 6000. Within two years, membership had reached 430 organizations scattered across thirty-six countries. In the early 1980s, membership was fairly divided among universities, governmental agencies, and companies. As with other user groups, its purpose was to share information concerning CDC's products and to relay requirements back to the manufacturer. VIM worked on software, database, hardware, programming standards, and applications. Two annual conferences were held where users exchanged information on their experiences with CDC's products. As with other such organizations, VIM shared software that ran on CDC's computers and produced newsletters and other relevant publications.

For further information, see: C. H. Warlick, "VIM," in Anthony Ralston and Edwin D. Reilly, Jr., eds., *Encyclopedia of Computer Science and Engineering* (New York: Van Nostrand Reinhold, 1983): 1559–1560.

X

XEROX CORPORATION. Xerox introduced copiers to the world and remains one of the leading producers of such equipment. It has also dabbled in data processing, primarily with office automation: typewriters, word processing equipment, and small departmental computers. The name of the company comes from the word *xerography*, the technology required to make dry copies of other documents. Such a process is also called electrophotography. The word *xerography* derives specifically from the Greek words for "dry" and "writing." The company shortened the term to Xerox to identify the specific equipment used in the process. The company brought out its first product in 1948, calling it a Xerox machine or copier and referring to the technology employed as xerography.

The technology at Xerox is the result of one man's work. On October 22, 1938, Chester Carlson (1906-) succeeded in copying another document. Carlson was born in Seattle, Washington, on February 8, 1906, grew up in San Bernardino, California, and while in high school worked for a printer, developing a strong interest in the graphic arts while at school, with chemistry. He studied chemistry and physics first at a junior college and then at the California Institute of Technology. Following graduation during the Great Depression of the 1930s, he obtained employment as a research engineer at Bell Telephone Laboratories†. He was laid off by Bell, worked for a patent attorney for a short while, and then joined an electronics firm called P. R. Mallory & Company. At night he studied for a law degree at the New York Law School.

His work with a patent attorney awoke him to the need for rapid, multiple copies of documents. Carlson therefore began searching for ways to make a device that could faithfully duplicate documents and settled on photoconductivity which involved having light strike a photoconductive material. In October 1937 he filed for a patent on his process. Between 1939 and 1944 over twenty companies turned down his overtures to manufacture copiers. Finally, the Battelle

Memorial Institute, a nonprofit research organization, signed a royalty-sharing agreement with Carlson to help refine his work. In 1947 the Haloid Company agreed to build xerographic machines. The first truly functioning office copier, the 914, did not appear until 1959. Previous devices were highly specialized and not intended for mass use in offices.

The 914 launched the company, thrusting it into a decade of prosperity since it was the only manufacturer of copiers then available. By the end of 1962 the company had shipped over 10,000 machines and generated a net income of $13.9 million. In 1961 Haloid Xerox became known simply as Xerox. By 1969 all copiers were called xerox machines, and anyone who wanted to make a copy of a document would "xerox" the paper, regardless of who manufactured the copier. The success of this technology made the Xerox Corporation a major supplier of office equipment. By 1970 many vendors realized that data processing suppliers would be invading the office in the years to come with their own products. Management at Xerox, with a large list of satisfied customers in that market, was of a similar mind.

In 1969 Xerox acquired an important manufacturer of computer-based products called Scientific Data Systems (SDS).† This company put Xerox into the data processing industry. At the time of the merger, Xerox acquired and managed 4,000 employees and assets of $113 million. In 1968 Xerox had logged in over $896 million in revenues, producing net income of $116 million. The data processing portion of the business was now known as Xerox Computer Services (XCS). In the 1970s it competed for business with small computers for such applications as accounting and as a replacement for work done with service bureaus.

Yet the acquisition was ill fated from the beginning. Revenues from XCS declined from $124 million in 1969 to $82 million in 1970 to $65 million in 1971. Its products were aging at the same time that the U.S. economy was experiencing a recession. The Sigma series of computers sold by XCS, especially the Models 3, 5, and 7, had all been announced in the 1960s and thus were becoming rapidly outdated and noncompetitive. Replacements appeared very late—in 1974 and early 1975—after other firms such as International Business Machines Corporation (IBM),† for example, had already brought out follow-on machines for their products of the 1960s. Furthermore, other vendors were already shipping newer products in quantity, costing Xerox lost sales and opportunities. Manufacturing difficulties in 1972–1973, coupled with growing manufacturing responsibilities to build copiers in plants originally assigned to construct computer-based products, made it difficult to deliver competitive equipment or in the volume necessary. Critical problems with disk drives in particular delayed the shipment of computer systems in the early 1970s. Two reorganizations in 1972 and 1973 did not help either, delaying the introduction of follow-on products.

Finally, in July 1975 Xerox announced that it would no longer build or sell computers. Other data processing products, however, which amounted to nearly

50 percent of its overall data processing revenues, continued to be sold. These did well and in 1979, for example, generated $475 million in revenues. Xerox Computer Services provided accounting and management applications for small and intermediate-sized customers much like a service bureau. The company built peripheral equipment and enhanced its position through acquisitions. One of the most notable of these acquisitions (made in 1972) was Diablo Systems, Inc., which made disk units, printers, and terminals. In 1975 Xerox bought Daconics, which made electrostatic printers and plotters. In December 1977 Shugart Associates joined Xerox, enhancing the company's ability to make flexible disk drives used with minicomputers, word processing equipment, and terminals. Centry Data Systems (CDS) with its expertise in the manufacture of rigid disk drives came into the fold in 1979.

The thrust of Xerox's data processing activities in the 1970s focused on office systems. It built word processors, copiers, and departmental systems. In 1977 it brought out the 850 and the Visual Type III, two important word processing systems for Xerox. At the same time the firm introduced the 9700 electronic printing system which allowed a user to take advantage of xerography to print 18,000 lines per minute from tape or a computer. In 1979 a user could add a tape drive to his or her printer. In 1978–1979 Xerox returned to the computer market with the 860, a system that did text editing, computation, and business applications. That same year it acquired WUI Inc., giving the company expertise in telecommunications. That year the company also announced Ethernet which allowed users to link together workstations, printers, and computers within a building. It became the basis of Xerox's telecommunications strategies and offerings during the early 1980s and competed directly against IBM's SNA and other communications systems from American Telephone & Telegraph (AT&T)† and Northern Telecom,† among others.

During the early 1980s Xerox continued to introduce copiers, large duplicators, and printing systems—some of which relied on laser technologies. It also competed with IBM, Sperry,† and others for sales of electronic typewriters. Additional networking products appeared to enhance the Ethernet local area network (LAN) and the company's word processing minicomputers.

Revenues in 1980 reached $7.886 billion and over the next three years leveled off between $8.2 and $8.3 billion annually. It closed 1984 with $8.7 billion. Net income, however, declined from a high of $598 million in 1981 to $291 million in 1984. It went from being the tenth largest data processing company in 1983 to the seventeenth in 1984, which still gave it a bigger presence in the data processing world than such firms as Texas Instruments† (thirty-second), Amdahl† (thirty-sixth), or Compaq Computer Corporation (forty-first). In short, it retained an important presence in the office products market.

For further information, see: Franklin M. Fisher et al., *IBM and the U.S. Data Processing Industry: An Economic History* (New York: Praeger Publishers, 1983); Gary Jacobs and John Hillkirk, *Xerox: American Samurai* (New York: Macmillan, 1986); Xerox Corporation, *1985 Fact Book* (Stanford, Conn.: Xerox Corporation, 1985) and its *The Story of Xerography* (Stamford, Conn.: Xerox Corporation, 1978).

Appendix A: Organizations Listed by Type

COMPANIES

Consultants

Gartner Group, Inc.
International Data Corporation (IDC)

Hardware Manufacturers

Amdahl Corporation
American Telephone and Telegraph (AT&T)
Apple Computer, Inc.
Atari Corporation
British Tabulating Machine Company (BTM)
Bull Company
Burroughs Corporation
Commodore International, Inc.
Computer Research Corporation (CRC)
Control Data Corporation (CDC)
Data General Corporation
Datapoint Corporation
Dataproducts Corporation
Digital Equipment Corporation (DEC)
Engineering Research Associates (ERA)
Fairchild Semiconductor Corporation
Ferranti, Ltd.
General Electric (GE)
Harris Corporation
Hewlett-Packard Company (H-P)
Honeywell Inc.
Intel Corporation
International Business Machines Corporation (IBM)
Itel
Kaypro Corporation
Memorex Corporation
National Cash Register Company (NCR)
Northern Telecom, Ltd. (NTL)
Northrop Aviation*
Osborne Computer Company
Philco
Powers Accounting Machine Corporation
Prime Computer, Inc.
Radio Corporation of America (RCA)
Raytheon Company
Remington Rand Corporation
Scientific Data Systems (SDS)
Sperry Rand Corporation
Tandy Corporation
Telex
Texas Instruments, Inc. (TI)
TRW
Verbatim Corporation
Viatron Computer Systems Corporation
Xerox Corporation

*Although not a manufacturer of computers, it has been included because it significantly influenced the nature of those systems in the late 1940s.

Retail Stores

ComputerLand Corporation
Radio Shack (*See* Tandy Corporation)

Service Bureaus

Automatic Data Processing (ADP)

Software Vendors

Computer Sciences Corporation (CSC)
Informatics, Inc.
Lotus Development Corporation
Microsoft Corporation

LABORATORIES, SCHOOLS, AND GOVERNMENT AGENCIES

Bell Laboratories
Bletchley Park
Lincoln Laboratory
Moore School of Electrical Engineering
National Bureau of Standards (NBS)
National Research Development Corporation (NRDC)
Office of Naval Research (ONR)

PROFESSIONAL ASSOCIATIONS

American Federation of Information Processing Societies (AFIPS)
American Society for Information Science (ASIS)
Association for Computing Machinery (ACM)
Association for Educational Data Systems (AEDS)
Association for Systems Management (ASM)
Charles Babbage Institute for the History of Information Processing (CBI)
CUBE
Data Processing Management Association (DPMA)
DECUS
GUIDE
Homebrew Computer Club
Institute of Electrical and Electronics Engineers, Inc. (IEEE)
Instrument Society of America (ISA)
International Federation for Information Processing (IFIP)
Joint Users Group (JUG)
National Computer Conference (NCC)
SHARE
Society for Computer Simulation (SCS)
Society of Information Display (SID)
USE
VIM

Appendix B: Chronology

1820	Thomas de Colmar invented a calculating machine and soon after began to sell versions of it called the Arithmometer
1875	Frank Baldwin began building calculators in Philadelphia
1880	William S. Burroughs established the Arithmometer Company, the precursor of the Burroughs Corporation
1882	National Cash Register Company (NCR) was established
1884	The Institute of Electrical and Electronics Engineers, Inc. (IEEE) was established
1885	Brunsviga Company introduced a popular Baldwin-Odhner calculator; Dorr E. Felt designed the Comptometer, a key-driven calculator that could add and divide
1886	Dorr E. Felt began manufacturing the Comptometer
1890	U.S. Census Bureau used card punch equipment developed by Herman Hollerith to tabulate the census
1892	General Electric Company (GE) was established; William S. Burroughs began marketing an adder-subtractor machine with the best printer available
1896	Thomas J. Watson, Sr. went to work for NCR as a salesman; Herman Hollerith formed the Tabulating Machine Company to sell punched card and tabulating products
1900–1910	Mechanical calculators became widely used in science
1902	Dalton Adding Machine Company came into existence
1904	Tabulator Ltd. was established in Great Britain to sell Hollerith's equipment
1907	Remington Accounting Machine Corporation was established
1911	Accounting and Tabulating Machine Corporation was established; Powers Accounting Machine Corporation was established to sell punched card equipment; Hollerith Tabulating Machine Company merged into the Computing-Tabulating-Recording Corporation (C-T-R), the precursor of IBM
1914	Thomas J. Watson, Sr. became general manager of C-T-R
1919	Radio Corporation of America (RCA) was established
1922	Bull Company was established in France to sell punched card equipment

1923	Moore School of Electrical Engineering was established at the University of Pennsylvania
1924	C-T-R was renamed the International Business Machines Corporation (IBM); Bell Labs was established to develop telephonic technologies and products
1927	Remington Rand was incorporated and immediately became a major supplier of office equipment; Remington Rand also purchased the Powers Accounting Machine Company
1930	The predecessor to Texas Instruments (TI) was established as the Geophysical Service
1934	IBM announced the IBM 405 Alphabetical Accounting Machine, the company's most popular product until the end of World War II; Federal Communications Commission (FCC) was created
1935	IBM announced the IBM 601 Multiplying Punch
1939	Hewlett-Packard Company (H-P) was established as a supplier of electronics
1945	ENIAC, the world's first electronic digital computer, became operational; IBM became the largest business machine vendor in the United States
1946–1959	First-generation computers predominated
1946	ENIAC was formally dedicated on February 16; J. Presper Eckert and John W. Mauchly resigned from the Moore School and in March established the Electronic Control Company to build and sell digital computers; Engineering Research Associates (ERA) was established with William Norris as president
1947	The Association for Computing Machinery (ACM) was established as the Eastern Association for Computing Machinery; Northrop Aviation signed a contract with the Electronic Control Company for the construction of the BINAC; Transistor was invented, making possible increased miniaturization of electronics and reduced costs for components during the 1950s and 1960s
1948	IBM introduced the Selective Sequence Electronic Calculator (SSEC), an electromechanical computer; American Telephone and Telegraph (AT&T) made public the development of the transistor on June 30; Electronic Control Company was reorganized into the Eckert-Mauchly Computer Corporation; Eckert-Mauchly and the U.S. Census Bureau signed a contract for the construction of the UNIVAC
1949	BINAC was delivered to Northrop Aviation; Data Processing Management Association (DPMA) was formed as the National Machine Accountants Association in Chicago
1950	Remington Rand bought the Eckert-Mauchly Computer Corporation; AT&T became the largest corporation in the United States; Computer Research Corporation (CRC) came into existence to build small computers for the U.S. military
1951	First UNIVAC was delivered to the U.S. Census Bureau; MIT established the Lincoln Laboratory, a major center for research on computer science; Bull Company introduced the Gama 2, a delay-line memory computer, for the French market
1952	In May IBM announced the IBM 701, the company's first commercially available computer; First IBM 701 was installed in December at IBM's

	corporate headquarters in New York; Society for Computer Simulation (SCS) was formed dedicated to the use of computer-based modeling; A UNIVAC was used to predict correctly the outcome of the U.S. presidential election, focusing considerable public attention on computers in general; Remington Rand acquired ERA; Thomas J. Watson, Jr. was elected president of IBM
1953	IBM shipped a 701 to the Los Alamos National Laboratory; ERA shipped its first 1103 computer; Kaypro Corporation was established; NCR introduced its first commercial computer, the CRC 102D; Remington Rand installed the first JOHNNIAC
1954	IBM began delivering the IBM 650, a medium-sized computer; TI announced that it has produced the first commercially available silicon transistors; GE bought the first UNIVAC dedicated to commercial applications
1955	First meeting was held of the IBM users' group, SHARE; IBM began designing STRETCH and shipping the IBM 702 computer; Four customers of UNIVAC formed a user's group, USE; Sperry Rand Corporation formed out of the merger of Remington Rand and Sperry Corporation; William B. Shockley formed his own company to manufacture semiconductors in Mountain View, California
1956	The IBM users' group, GUIDE, was formed; IBM began delivering the IBM 705 computer; GE announced the Electronic Recording Method of Accounting (ERMA) computer; RCA announced the BIZMAC computer for commercial applications
1957	IBM announced the IBM 709 and began shipping the first disk drive, the IBM 350, as well as the IBM 650 RAMAC system and the IBM 305; IBM also exceeded $1 billion in sales and introduced FORTRAN, one of the earliest high-level programming languages; John Burns became chief executive officer at RCA just after having been a consultant to IBM on that firm's future marketing and product plans; UNIVAC II was introduced; Digital Equipment Corporation (DEC) came into existence to sell minicomputers; Fairchild Semiconductor Corporation was established as one of the first manufacturers of chips and other electronic components; Electrodata introduced and shipped the DATATRON 220, a vacuum tube computer; Control Data Corporation (CDC) was established and later announced its first computer, the 1604 system, a solid-state device
1958	IBM introduced second-generation computers with the IBM 7070 and the IBM 7090 and began shipping the IBM 709; RCA began marketing commercial versions of BIZMAC; Philco began delivering the transistorized TRANSAC S–2000, one of the first commercially available computers based on transistors; FLOW-MATIC, a business programming language, became available for use on the UNIVAC; First AN/FSQ7 computer, part of the SAGE system, became operational at McGuire Air Force Base in New Jersey
1959–1964	Second-generation computers predominated
1959	Computer Sciences Corporation (CSC) was established as one of the first companies created to develop software under contract; Robert N. Noyce of Fairchild Semiconductor applied for a patent on the chip; IBM announced the IBM 1401 and also the IBM 1620, an early, low-cost, small scientific computer; IBM and American Airlines signed an agreement jointly to

	develop the SABRE airline reservations system; First ERA 1101 was delivered to the U.S. Bureau of the Census; PEGASUS series of British computers became available; First office copier, called the 914, was introduced by the Haloid Company
1960	International Federation for Information Processing (IFIP) was formally established; CDC began shipping its first computers, the 1604 and the 160; RCA announced the 601, a large general-purpose computer; DEC introduced the PDP–1, the company's first minicomputer; IBM began shipping the 1400 system
1961	Memorex Corporation was established; American Federation of Information Processing Societies (AFIPS) was formed; IBM established the Spread committee, which recommended production of the S/360; GE announced the GE 225, intended for both commercial and scientific processing; Haloid Xerox became known as Xerox Corporation; Scientific Data Systems (SDS) was established; MIT created the first time-sharing system; DEC users formed the Digital Equipment Computer Users Society (DECUS); TI constructed the first computer based on the integrated circuit
1962	Informatics, Inc. was established to develop and sell software and programming services; Dataproducts Corporation was established for the purpose of manufacturing peripheral equipment; SDS delivered its first computer, the SDS 910; The PDP–4 minicomputer was first shipped; CDC announced its first large computer, the 6600; Telex acquired Midwestern Instruments, an electromechanical equipment manufacturer that made magnetic tape drives; Society for Information Display (SID) was formed; The National Machine Accountants Association changed its name to the Data Processing Management Association (DPMA); Burroughs users formed CUBE to share information and experiences
1963	Tandy Corporation gained control of Radio Shack; Ferranti Ltd., a major British electronics firm, stopped manufacturing computers; Burroughs Corporation introduced its popular B5000, the firms's largest computer of the 1960s; DEC announced a new minicomputer; Bell Punch Company of Great Britain began selling electronic calculators built out of discrete components
1964–1970	Third-generation computers, hallmarked by the IBM S/360, were predominant
1964	On April 7, in the largest product announcement to date in the data processing industry, IBM announced the S/360, the first major family of computers, consisting of over 150 different products—it became the most successful product in American economic history; CDC shipped its first 6600; RCA announced the Spectra series of computers to compete against IBM's S/360; Phrase ''database'' first appeared in U.S. military reports on managing machine-readable files; International Data Corporation (IDC) was formed; SDS introduced its first computer to use monolithic integrated circuits, the SDS 92
1965	DEC shipped its first minicomputer made with integrated circuits, the PDP–8; IBM began shipping S/360 computers
1966	H-P announced its first computer, the Instrumentation Computer; IBM introduced its database manager, DL/I

1967 — Memorex introduced its initial disk packs; Itel was established as SSI Computer Corporation

1968 — Itel signed its first computer lease; Data General Corporation was established; Computer Terminal Corporation, later to be known as Datapoint Corporation, was formed; CDC bought Commercial Credit, entering the financial market for the first time; First microcomputer was announced in October by Viatron Computer Systems Corporation; NCR announced the Century Series of computers; Intel was formed to manufacture integrated circuits and later introduced its first product, the 1K random-access memory (RAM)—the first in the world

1969 — U.S. Justice Department filed an antitrust suit against IBM, launching one of the longest legal cases in American history; Greyhound Corporation also filed an antitrust suit against IBM; Data General began shipping its first minicomputer, the Nova; Largest manufacturer of floppy disks, Verbatim Corporation, was established; Xerox Corporation acquired SDS

1970 to present — Fourth-generation computers were predominant

1970 — Users of CDC equipment formed VIM; CDC attains $1 billion in sales; GE withdrew from the computer market and sold its operations to Honeywell, thereby jointly forming Honeywell Information Systems (HIS); IBM announced the S/370 family of computers to replace its highly successful S/360 and introduced the 3420 tape drive, the industry standard for nearly the next two decades; Sperry announced the 1110, a general-purpose computer; IDMS, a popular database management program of the 1970s, was announced by Cullinane Corporation; Amdahl Corporation was established to build IBM-compatible computers

1971 — CDC announced the CYBER 70 family of computers; RCA announced its withdrawal from computer manufacturing, creating a sensation in the data processing industry; Memorex announced its 3670 disk drive to compete against IBM's 3330 disk system; Telex began shipping IBM 3420-compatible tape drives, the 6420; Intel developed the microprocessor and introduced the 4004 chip; Prime Computer, Inc. was established; IBM began shipping its IBM 3330 disk drive, the industry standard throughout the 1970s; Mass-produced pocket calculators became available in the United States

1972 — H-P announced its first hand-held calculator, called the "electronic slide rule," or the HP–35; Keuffel and Esser, a major supplier of slide rules, ceased to manufacture the device; TI announced the 8-bit chip and introduced the DataMath, an early hand-held calculator; Intel introduced the 8008 microprocessor, an 8-bit chip; Telex filed a suit against IBM for monopolistic behavior—it became a sensational case during the 1970s; FCC ruled that the U.S. government should not regulate the data processing industry; Prime Computer introduced its first product, the Prime 200 minicomputer; Atari Corporation—the first company to produce computer-based video games—was formed to market "Pong" the first video game

1973 — IBM filed a suit against Telex for patent infringement; CDC and IBM ended their lawsuit, with the result being that IBM's Service Bureau Corporation (SBC) became a part of CDC; The ENIAC patent was invalidated by a U.S. court, ending a lengthy legal battle; First National Computer

	Conference (NCC) was held in New York; IBM first used MOSFET memory in the S/370 Model 158 and Model 168; Frank Cary was elected chief executive officer of IBM; Most computer vendors sold products based on the integrated circuit
1974	U.S. government filed an antitrust suit against AT&T, proposing the breakup of the telephone company; IBM announced MVS, an operating system for large computers; TI introduced a series of hand-held computers, including the SR–50 and SR–51; An article published in *Radio Electronics* on how to build a "personal computer" stimulated the interest of many individuals who went on to form companies to build just such devices
1975	JUG, a DEC user club, held its last meeting after fourteen years of existence; Homebrew Computer Club came into existence as the Amateur Computer Users Group; Xerox Corporation withdrew from the manufacturing and marketing of computers; Microsoft Corporation was established; TI introduced the first digital watch; AT&T announced both its Dataspeed 50 terminals to compete against IBM's 3270s and the Dimension Private Branch Exchange (PBX) system; Over 1,000 firms were operating in the data processing industry marketing programs and software; Altair computer was introduced in *Popular Electronics*, the first commercially available personal computer of the 1970s
1976	ComputerLand opened its first retail store to sell computer products
1977	Apple Computer, Inc. was formally created, and what would become Apple's most successful product, the Apple II personal computer, was introduced; Telex acquired Gulliver Technology, which manufactured a 6,250 bits-per-inch tape drive; IBM announced its 303X family of large computers; Tandy Corporation announced its first microcomputer, the TRS–80, which it sold through its Radio Shack retail outlets; Xerox introduced the 850 and the Visual Type III, two important word processing systems; TI introduced the TI–58 and TI–59 hand-held programmable calculators; Charles Babbage Institute was established to study the history of Information processing
1978	IBM announced the IBM 8100 distributed processors
1979	IBM announced the IBM 4300 family of medium-sized computers; Tandy announced the TRS–80 Model II, a state-of-the-art microcomputer
1980	FCC issued Computer Inquiry II, announcing it would henceforth only regulate essential transmission services that moved data and voice; IBM announced the IBM 3380 disk drive, the most advanced in the industry; Gartner Group, Inc., a leading U.S. consultant on data processing industry trends, was created by Gideon I. Gartner
1981	Osborne 1 microcomputer was introduced at the West Coast Computer Faire; IBM introduced its Personal Computer (PC)
1982	AT&T and the U.S. government came to an agreement leading to the breakup of the Bell System into AT&T and a series of regional companies; U.S. government dropped its thirteen-year attempt to sue IBM for antitrust practices due to a lack of evidence; Lotus Development Corporation was formally established; Lotus 1–2–3 was announced, one of the most widely used spread-sheet software packages for microcomputers; Kaypro introduced a minicomputer, the Kaycomp II; Pac Man, a popular video game, was introduced

1983	American Bell (Baby Bell) was established; First major fiber-optics line was opened by AT&T between Washington, D.C., and New York; Osborne Computer Corporation declared bankruptcy; IBM introduced the PCjr; TI lost $150 million marketing personal computers; Shipments of microcomputers exceeded 1.5 million
1984	AT&T formally broke up into AT&T and regional companies; Apple Computer introduced the Macintosh microcomputer; IBM introduced the one-million-bit RAM
1985	IBM announced the 3090 large MVS/XA computers; Lotus Development Corporation became the world's largest supplier of independent software
1986	Burroughs and Sperry announce their merger to form Unisys Corporation; Widespread layoff of employees by various computer vendors continued due to severe slump of all sales in the data processing industry
1987	IBM first shipped the IBM 9370 mid-range computer

Index

Boldface numbers indicate the location of the main entries in the text.

About the Author

JAMES W. CORTADA is Senior Marketing Programs Administrator for the IBM Corporation. He is the author of several books on the history and management of data processing, including *EDP Costs and Charges*, *Managing DP Hardware*, and *An Annotated Bibliography on the History of Data Processing* (Greenwood Press, 1983), as well as two forthcoming companion volumes to the *Historical Dictionary of Data Processing: Organizations* which will cover technology and biographies in the history of data processing. Dr. Cortada has also published numerous articles in a variety of journals.